T0220300

Cambridge Elements ≡

Elements in the Philosophy of Biology
edited by
Grant Ramsey
KU Leuven
Michael Ruse
Florida State University

PALEOAESTHETICS AND THE PRACTICE OF PALEONTOLOGY

Derek D. Turner
Connecticut College

CAMBRIDGE
UNIVERSITY PRESS

CAMBRIDGE
UNIVERSITY PRESS

University Printing House, Cambridge CB2 8BS, United Kingdom

One Liberty Plaza, 20th Floor, New York, NY 10006, USA

477 Williamstown Road, Port Melbourne, VIC 3207, Australia

314–321, 3rd Floor, Plot 3, Splendor Forum, Jasola District Centre,
New Delhi – 110025, India

79 Anson Road, #06–04/06, Singapore 079906

Cambridge University Press is part of the University of Cambridge.

It furthers the University's mission by disseminating knowledge in the pursuit of
education, learning, and research at the highest international levels of excellence.

www.cambridge.org
Information on this title: www.cambridge.org/9781108727822
DOI: 10.1017/9781108671996

First published 2019

A catalogue record for this publication is available from the British Library.

ISBN 978-1-108-72782-2 Paperback
ISSN 2515-1126 (online)
ISSN 2515-1118 (print)

Paleoaesthetics and the Practice of Paleontology

Elements in the Philosophy of Biology

DOI: 10.1017/9781108671996
First published online: September 2019

Derek D. Turner
Connecticut College

Author for correspondence: Derek D. Turner, derek.turner@conncoll.edu

Abstract: The practice of paleontology has an aesthetic as well as an epistemic dimension. Paleontology has distinctively aesthetic aims, such as cultivating a sense of place and developing a better aesthetic appreciation of fossils. Scientific cognitivists in environmental aesthetics argue that scientific knowledge deepens and enhances our appreciation of nature. Drawing on that tradition, this Element argues that knowledge of something's history makes a difference to how we engage with it aesthetically. This means that investigation of the deep past can contribute to aesthetic aims. Aesthetic engagement with fossils and landscapes is also crucial to explaining paleontology's epistemic successes.

Keywords: paleontology, aesthetics, scientific practice, scientific realism, dinosaurs

ISBNs: 9781108727822 (PB), 9781108671996 (OC)
ISSNs: 2515-1126 (online), 2515-1118 (print)

Contents

1 Introduction

We philosophers of science tend to privilege epistemological questions – that is, questions about scientific knowledge, evidence, confirmation, and related notions. Researchers working on the philosophy of paleontology, or on the historical sciences more broadly, have generally continued in this vein. My goal is to push back against this bias toward the epistemic, and to do so by showing that paleontological research is a form of aesthetic engagement with fossils and with landscapes. If this is right, then much recent philosophical discussion of paleontology has been too narrow in focus.[1] Over the last fifteen or twenty years, philosophers of science from Carol Cleland (2002) to Adrian Currie (2018) have made a great deal of progress in understanding how paleontologists reconstruct the history of life on Earth.[2] But with the exception of Caitlin Wylie's (2009, 2015) research on the practice of fossil preparation, and Currie's (2017b) argument that we should see "paleoart as science," nearly all of this work on the epistemology of historical science has neglected the aesthetic dimensions of the practice of paleontology.[3] Yet we cannot really understand the science without thinking about those aesthetic dimensions. I'll use the term *paleoaesthetics* to refer to the study of the aesthetic dimensions of historical science.

In making this argument, I draw heavily upon work in environmental aesthetics. Some recent work in environmental aesthetics has obvious relevance to philosophy of science. Sadly, due to academic specialization, philosophers of science have not engaged with views from environmental aesthetics that could have much to teach us about science.

My focus here is on paleontology – or the paleosciences, more broadly construed to include some earth science and evolutionary biology – in part because paleontology is one area of science where the aesthetic dimension is especially easy to see. Although I won't try to make the case here, I strongly suspect that at least some of what I say about the aesthetic dimensions of paleontology will apply to other areas of science.

We philosophers of science have not done a very good job in engaging with the work of scholars who have analyzed the place of dinosaurs in the larger

---

[1] This critique also targets some of my own work. For example, in my (Turner 2011) and (Turner 2014) works, I tried to offer road maps to some of the main issues in the philosophy of paleontology, but I didn't discuss aesthetic considerations at all.

[2] Contributions to this literature on the epistemology of historical science include Chapman and Wylie (2016); Cleland (2002, 2011); Currie (2015, 2017a, 2018); Currie and Turner (2016); Forber and Griffith (2011); Jeffares (2008); Kosso (2001); Parsons (2001); Tucker (2004, 2011); Turner (2007, 2009).

[3] Currie (2017b) is also inspired partly by Witton, Naish, and Conway's (2014) reflections on paleoart.

culture (Boym 2002, ch. 3; Mitchell 1998). These scholars are not really focusing on paleontological science, and yet their work is very relevant to understanding the larger cultural context in which that science is occurring. Paleoaesthetics might give us a way of thinking about the science that makes some of these other cultural connections easier to draw, though I will not be able to do much more than hint at these connections in Section 9.

In recent years, there has been a lot of work on non-epistemic values in science, especially following Heather Douglas's (2000, 2009) groundbreaking work on inductive risk. Inductive risk refers to the risk of either accepting a false hypothesis (a false positive), or rejecting a true one (a false negative). In different contexts, we might think that one type of error would be far less tolerable than another. For example, in a toxicological study looking at the health effects of a particular synthetic chemical, you might think that a false negative (wrongly concluding that a dangerous chemical is safe) would be far worse than a false positive. A false negative means that people could suffer. Douglas argued that we cannot really handle inductive risk without allowing our ethical, social, and political values to inform the way that scientific research is done. Due to historically contingent facts about how the issues got framed in the literature, most people today writing about non-epistemic values in science are thinking about inductive risk, and about ethical, social, and/or political values. Douglas (2009) and Elliott (2017) scarcely mention aesthetic value.

It might seem like paleontology is the very last place that one should look in order to find non-epistemic values implicit in the practice of science. Paleontology is about as far removed from any policy issues as one can get, the paradigm case of a policy-irrelevant science where the costs of getting something wrong are rather low.[4] If we're wrong about whether *Torosaurus* and *Triceratops* were really different genera, no one's health or wellbeing will be affected, although some dinosaur aficionados no doubt have some emotional investment in the issue (Scannella and Horner 2010). Nevertheless, if we shift the focus to aesthetic values, it turns out that paleontology is an excellent test case for thinking about non-epistemic values in science. I argue that non-epistemic (aesthetic) values are woven into the practice of paleontology. As Elizabeth Anderson has argued, "science is value free if and only if values are science free" (Anderson 2004, p. 7). The aesthetic values of fossils and landscapes are not science free.

Given the brief and programmatic nature of this Element, I offer a high-altitude view of things. I will deal breezily with some issues about which there is

---

[4] It's not, however, totally irrelevant to policy. I've argued elsewhere that paleontology might have insights to contribute to conservation policy (Turner 2016). See also Dietl and Flessa (2011).

already a large literature, such as the scientific realism debate, or the literature on values in science. Some of the arguments will no doubt need to be spelled out with greater care and attention to possible objections. Implicit in this high-altitude picture is a critique of much recent philosophy of science, a critique that I bring into focus in Section 6, which engages with what has traditionally been the central concern of scientific realism – namely, how best to explain the success of natural science. There my longstanding skepticism about scientific realism (see Turner 2007, 2018) will take a somewhat new shape.

To give a sense of what's to come, here is a brief overview of some of the claims I develop:

- Paleontology has aesthetic as well as epistemic aims, including cultivating sense of place and gaining a deeper aesthetic appreciation of fossils.
- Learning about something's history makes us better appreciators of its aesthetic qualities.
- Fossils and landscapes have transformative aesthetic value, insofar as they have the capacity to transform our aesthetic preferences.
- Functional morphology (the investigation of the functions of fossilized structures) is a form of aesthetic inquiry.
- Some aspects of the practice of paleontology (fossil preparation, 3D printing fossils, paleoart, making field sketches) are artistic practices. The epistemic successes of the paleosciences depend crucially on these practices.
- Entrenched textual metaphors for fossils (the fossil "record") make the aesthetic dimensions of paleontology harder to see, but those metaphors are optional.
- Sometimes, in dinosaur science, high-profile scientific debates come down to different researchers' aesthetic engagement with a single fossil specimen.

I'll conclude, in Section 9, with some reflection on a case where recalcitrant aesthetic biases seem to get in the way of scientific investigation.

This project falls squarely in the tradition of *practice-oriented* philosophy of science, as contrasted perhaps with more theory-centric approaches. Thus, this Element may complement and serve as something of a dialectical sequel to Turner (2011), which explored paleontologists' contributions to macroevolutionary theory. Theory-oriented and practice-oriented approaches are fruitfully pursued in tandem, so what follows is best thought of as a practice-oriented expansion of the philosophy of paleontology.

There have been many efforts to articulate what a practice-oriented approach to science might look like, with some differences of emphasis (Chang 2014; Hacking 1983; Rouse 2002, 2015; Waters 2014; Woody 2014; Wylie 2002). Practice-oriented work in the philosophy of science also tends to be more open to inputs from science and technology studies (STS) and social studies of science. Perhaps one helpful way to understand the shift to a more practice-oriented philosophy of science is to focus on the process/product distinction: Much traditional philosophy of science analyzes the products of scientific research – theories, for example. The guiding questions are all about the structures of theories, their status, their relationship to the evidence and to each other, what attitude we should have toward them, and so on. Advocates for the practice turn in the philosophy of science seek to shift the focus to the processes of scientific research. What are the various things that scientists are doing? And what should they be doing? That includes practices of theory construction and testing, but many more things besides.[5] Paleontology includes all sorts of different practices: fossil collection, interpretation of field sites, fossil preparation, statistical analysis, curation of exhibits, biomechanical modeling, database construction, taphonomic experiments, coding characters, paleoart, the use of technology such as CT scanners, and much else. Rather than trying to shoehorn all of these various activities into hypothesis- or theory-testing, we might do better to understand these practices on their own terms. Most do have some connection to empirical testing, but there may also be more going on than that.

Part of the agenda of this Element is to show what a more practice-oriented philosophy of paleontology would look like. This project might seem to sit a little uneasily with some of the most exciting work in the history and philosophy of paleontology that's been done in recent years. Much of that work has focused on the "paleobiological revolution" of the 1970s and 1980s, an era when a new generation of paleontologists sought to show that paleontology has much to contribute to the theory of evolution at larger scales. Historian of science David Sepkoski (2012) has magnificently documented this important episode (see also the papers collected in Sepkoski and Ruse 2009). Given the importance of this bid for theoretical relevance, it's understandable that some work in the philosophy of paleontology should be more theory-centric, focusing on ideas like punctuated equilibria, species selection, evolutionary contingency, and the study of macroevolutionary trends. In pursuing a more practice-oriented approach here, my aim is not to suggest that there is anything wrong with focusing, say, on macroevolutionary theory. Think of this more as an effort to expand the philosophy of paleontology by examining some aspects of the practice that have not yet gotten sufficient attention, and by correcting for some existing philosophical biases in favor of epistemological questions and in favor of conceptual and theoretical analysis.

---

[5] I thank Adrian Currie (personal communication) for suggesting this application of the process/ product distinction.

The claim that the paleosciences have a neglected aesthetic dimension would seem to presuppose a fairly sharp distinction between epistemic and non-epistemic (e.g. aesthetic) values. Some philosophers of science have questioned that distinction (Rooney 1992). Other philosophers have defended it (Steel 2010). This suggests two different ways of thinking about how the aesthetic and epistemic dimensions of science are related. According to the first approach, which we might call the "intertwining view," the aesthetic and epistemic dimensions of paleoscience remain distinct, though they weave together in productive and mutually supportive ways. Those who defend the epistemic/non-epistemic distinction will naturally incline toward the intertwining view. According to the second possibility, which I will call the "blurring view," there is no clear distinction between the epistemic and the aesthetic dimensions of scientific practice. Philosophers who are skeptical about the epistemic/non-epistemic distinction might feel more drawn to the blurring view. The difference between the intertwining view and the blurring view is the difference between saying that knowledge of the deep past enhances aesthetic engagement, and saying that investigating the deep past is just a form of aesthetic engagement.

I myself am deeply sympathetic to the more radical blurring view. Once we really absorb the lessons of paleoaesthetics, it may turn out that there is no sharp distinction between it and paleoepistemology. In a couple of places in what follows, I will argue that the blurring view can help address potential worries and objections. For most purposes, though, I am also perfectly happy to work with the more modest intertwining view. Going forward, I will sometimes be a little fast and loose with respect to the distinction between these two views, since the difference between them does not matter much for the main argument. My primary goal is to show that the paleosciences have an aesthetic dimension. Working out whether the blurring *vs.* the intertwining view best captures what's going on is of secondary importance.

Finally, the argument of the Element runs in two directions. One direction goes from the epistemic dimensions of paleontology to the aesthetic, in an effort to show that investigating the past contributes to (or blurs into) aesthetic engagement. In later sections, the argument will cut back in the other direction, in an effort to show how practices that are clearly aesthetic contribute to (or blur into) epistemic or investigative activity.

2 Paleoaesthetics

Western Canada's Dinosaur Provincial Park is one of those special places where you can actually see dinosaur bones lying around underfoot. Up above you, the rolling prairies of eastern Alberta extend seemingly indefinitely in all directions – endless wheat fields and ranches. If you look to the north, you can imagine the wind coming down from the oil sands of far northern Alberta, and beyond that, the Arctic. Drive west, and you'll eventually see the Canadian Rockies rising up from the prairie, beyond Calgary. Where the Red Deer River cuts through the prairie on its way to meet the Saskatchewan River, which eventually flows east into Hudson's Bay, it has created a wide swath of arid badlands, sparsely vegetated erosional hills and gullies and slots that are basically a kind of northern desert. This was the site of Canada's great dinosaur rush in the early twentieth century, when Barnum Brown in his famous fur coat, along with his crew from the American Museum of Natural History, flatboated down the Red Deer River and quarried out huge quantities of dinosaur bones. But there is also more here than just dinosaurs.

There are places in the badlands where you can see easily recognizable dinosaur bones (usually from hadrosaurs) just eroding out of the ground. But the most remarkable thing I saw on a recent visit there was something that contributed even more than the dinosaur remains to my getting a sense for the place, and for its past: an exquisitely preserved 70 million-year-old freshwater clam bed.

There were clamshell fragments lying around on the ground that looked like they could have been deposited last year or even just last week. Nearby, at a place where the land had eroded out in a way that afforded a profile view, you

Figure 1 Dinosaur (probably *Hadrosaur*) bone, Dinosaur Provincial Park, Alberta, Canada. Photo by the author.

Figure 2 Freshwater clams, Dinosaur Provincial Park, Alberta, Canada. Photo by the author.

could see that the clam bed formed an entire stratum, so that when we were looking at the clamshells lying on the ground, we were actually standing on top of a clam bed layer. Seventy million years ago, during the late Cretaceous, we would have been standing in a green coastal plain, with wide rivers meandering their way out to the inland seaway that bisected North America at the time. The coastline would have been just to the east. The prehistoric clams flourished in bends of these rivers, whose sediments helped create the rock formations that the Red Deer River, in its turn, has slowly eaten its way through, such as the Dinosaur Park Formation. This historical knowledge changes one's whole experience of the landscape, giving that experience a stereotemporal quality. Like a stereoscope that uses two images of the same item to create an impression of depth, stereotemporal experience merges our current perceptual experience of the landscape with our understanding of its history. There are the disorienting arid badlands around you in the moment, with the river not too far away, in some forgotten direction, and the prairie up above. And there is the green coastal plain with heavy forests and wide rivers. If you read paleontologists' popular writings about their work, you can often find them musing about this stereotemporal experience, in which they are engaging aesthetically with a landscape in the here and now by contemplating a deep prehistory that the current landscape sometimes only hints at, but sometimes sings about.

Figure 3 Freshwater clam bed, seen in profile, Dinosaur Provincial Park, Alberta, Canada. Photo by the author.

Keith Basso (1996), an anthropologist, has written movingly about how people develop a sense of place in and by engaging with landscapes. The cultivation of sense of place involves a two-way interaction. We humans modify the landscape and assign meanings to it. Dinosaur Park, for example, has a complex history of human use that includes Barnum Brown's surprisingly massive quarry operations – an extractive enterprise that has much in common with other kinds of mining and exploitation of natural resources (Rieppel 2015). This was the scene of a historic "dinosaur rush," akin to a gold rush. The term "badlands" itself reflects the agricultural and pastoral priorities of Euro-Canadian settlers, but First Nations people have also lived in and modified the prairie landscape for thousands of years. Places also do things to us. It is tempting to attribute some agency to the badlands, even if it's only metaphorical. We might say that the barren landscape is actively resisting development, or that it has actively changed culture and science by yielding up all those dinosaur fossils. Certainly the landscape has an effect on you when you go there. (As an exercise, think about a landscape that has particular significance for you, one that has affected your life in some way.) We cultivate sense of place by experiencing this give-and-take between landscapes and human culture. Basso writes that "when places are actively sensed, the physical landscape becomes wedded to the landscape of the mind" (Basso 1996, p. 55). Physical landscapes shape our thinking and our sense of who we are, even as we shape them through use—or, in the case of paleontology, through collection and study. And our sense of a place has much to do with narratives, often including scientific narratives, about what transpired there.[6]

It is tempting to talk about aesthetic engagement with fossils and landscapes as if these were different things, and yet the clamshells and other fossils found *in situ* serve as reminders that fossils actually belong to landscapes and places. Even once collected, prepared, and displayed, the fossil retains a historical connection to the landscape. We routinely name fossils in ways that evoke the places they come from. *Albertosaurus* and *Edmontosaurus* are recognizable western Canadian dinosaurs. The provenance of collected and prepared fossils also links them back to landscapes. Borrowing an expression from archaeologist Richard Bradley, we might say that fossils are "pieces of places," so that engaging with a fossil is a form of indirect engagement with place (Bradley 2000, p. 96). Seeing a dinosaur skeleton in a museum can connect us both with a distant place and with the deeper history of that place.

The clamshells are doubly significant because while they point to a prehistoric landscape that could scarcely have been more different from today's eastern Alberta, they also hint at continuity and stability. Freshwater clams and mussels living today are barely distinguishable, morphologically, from the Cretaceous ones. The clamshells that we saw looked like they could have been deposited last week. But as it happens, in other places, freshwater clams are still thriving and doing their thing. This is a remarkable case of longer-term evolutionary stasis in the face of radical ecological and geological changes over the past 70 million years. The fossilized clams speak of a continuity between the late Cretaceous and the present, and they render the former just a bit less alien, while adding a new layer of puzzlement to the relationship between past and present. Why do some things change so little, while other things change so much? Knowing that 70 million years ago, this was a bend in a slow river, not too far up from the sea, diminishes the uncanniness of finding clamshells in the badlands, but it also opens up new things to wonder about.

---

[6] There has been a lot of interest lately in historical narrative explanation (Beatty 2016; Currie and Sterelny 2017). In addition to doing explanatory work – which is what philosophers of science tend to focus on – narratives may also contribute to the aesthetic goal of cultivating sense of place.

2.1 How Scientific Investigation Contributes to Aesthetic Engagement

Philosophers of science interested in paleontology and the earth sciences would do well to engage with relevant work in environmental philosophy. One of the leading views in environmental aesthetics, first developed by Allen Carlson in the 1970s and 1980s, is known as *scientific cognitivism*. The rough idea is that empirical knowledge – and especially, though perhaps not only, scientific knowledge – enhances our aesthetic engagement with nature and makes us better appreciators of plants, animals, and natural environments (Carlson 1977, 1981, 2000, 2009; Matthews 2008; Parsons 2008b; Saito 2008). Scientific cognitivists in environmental aesthetics may also draw some inspiration from Aldo Leopold's classic essay, "The Land Ethic" (in Leopold 1989). Leopold treats the beauty of the land community as one of several anchoring environmental values, and implicit in much of his writing is the idea that careful study of nature can help us to appreciate that beauty. Scientific cognitivism is controversial, and others working in environmental aesthetics have developed rival accounts of aesthetic engagement with nature that do not emphasize knowledge so heavily (Brady 2003; Berleant 1995). Here I defend a view inspired by Carlson's that I'll call *historical cognitivism*. According to historical cognitivism, knowing the history of something—whether a fossil, or a landscape, or anything else—deepens and enhances one's aesthetic engagement with that thing, and helps one to better appreciate its aesthetic qualities.

The most straightforward argument for historical cognitivism involves a simple thought experiment. In a paper called "Faking Nature" that has become a classic of environmental philosophy, Robert Elliot argues that "the manner of a landscape's genesis . . . has a legitimate role in determining its value" (Elliot 1982, p. 383). Elliot then uses a thought experiment to drive the point home:

> [I]magine I have been promised a Vermeer for my birthday. The day arrives and I am given a painting which looks just like a Vermeer. I am understandably pleased. However, my pleasure does not last for long. I am told that the painting is not a Vermeer but instead an exact replica of one previously destroyed. Any attempt to allay my disappointment by insisting that there just is no difference between the replica and the original misses the mark completely (pp. 383–4).

Two paintings can be qualitatively identical, but if they have different histories, we might justifiably value them differently.[7] In Elliot's example, knowing the historical genesis of the painting deepens our aesthetic appreciation of it. If you

---

[7] This is, to some degree, an appeal to intuition, and not all philosophers agree about the aesthetic relevance of history (see, e.g. Sandis 2016). But Sandis focuses mainly on forgeries and replicas.

have false beliefs about that painting – for example, if you think it's a Vermeer, when in fact it is a perfect forgery – then your aesthetic engagement with the artwork is misfiring in at least one important way.

Lukas Rieppel (2016) has pointed out that different standards of authenticity apply in the art world *vs.* the world of natural history museums. Many fossils on display in museums are in fact casts, or high-fidelity replicas. So the analogy between artworks and fossils isn't perfect. But Rieppel's point just further underscores the need for a cognitivist approach. If we do think that there is an aesthetic difference between genuine fossils and casts, that difference is going to turn out to be a matter of history. The historical cognitivist might still value replicas highly, depending on the context; the point is that the replica's history is relevant to *how* we should value it.[8]

Carolyn Korsmeyer (2012, 2016) goes a bit further than Elliot does in trying to explain what underwrites our intuitions about forgery cases. Korsmeyer agrees that "a replica does not inspire the same admiring attention as a real thing, even when it is perceptually indiscernible from an original," and she concludes that an object's aesthetic qualities are not exhausted by its perceptible qualities (Korsmeyer 2016, p. 220). But why should historical genuineness, as she calls it, make any difference to aesthetic value? Her answer is that interaction with old objects places us into causal connection with the past. Having a genuine Vermeer on your wall places you (in some sense) into contact with Vermeer himself, and that connection to something or someone in the past is, Korsmeyer argues, what we really care about. Korsmeyer's account may be part of the story about how history matters to the aesthetic appreciation of fossils and landscapes.[9]

Simon James (2015) ties the value of old things to narrative: "old inanimate objects typically deserve special treatment, not simply because they are old, but because they embody – or in some other way relate to – narratives that humble people will take seriously in their practical deliberations" (James 2015, p. 329). On James's view, what matters, say, in the case of the authentic Vermeer, is its place in a larger historical narrative about the history of European art. There are some differences of emphasis between James's and Korsmeyer's accounts. Korsmeyer places a lot of weight on the sense of touch as being the thing that connects us most directly and viscerally to the past—think of the experience of picking up a piece of petrified wood from the late Cretaceous. James, on the

---

In the next section, I'll describe a non-forgery case where it just seems really hard to deny that historical knowledge bears on aesthetic appreciation.

[8] Erich Hatala Matthes (2017, 2018) has also argued that replicas can have multiple purposes, and so it is not always reasonable to complain that replicas are "inauthentic."

[9] But see also Matthes (2018) for an interesting critical response to Korsmeyer.

other hand, focuses more on how virtuously humble people would respond to certain sorts of historical narratives. For present purposes, though, it's more important to note the similarity in their two approaches. Although they do not use this terminology, both Korsmeyer and James seem committed to versions of historical cognitivism; both think that our knowledge of history matters profoundly to how we engage with objects in the present.

Historical cognitivism implies that knowledge about the deep past enriches our aesthetic engagement with fossils and landscapes. Knowing that the freshwater clam fossils are 70 million years old, and that the clam bed was once at the bend of a river flowing east across a lush coastal plain, changes your experience of the fossils and the badlands. The geo-historical knowledge connects you to the place. On the one hand, scientific investigation is a structured epistemic activity, with epistemic goals, methods, and tools. But scientific investigation also has distinctively aesthetic payoff, contributing to sense of place, or contributing to our appreciation of fossils.

One might agree with much of what I've said so far, and still sympathize with an interesting view that I will call the *accidental aesthetic payoff* view. According to this view, scientific investigation in paleontology and related fields aims narrowly at knowledge. Knowledge of the history of fossils and landscapes does enhance our engagement with those things, but that is a merely accidental aesthetic payoff. Scientific investigation is still, first and foremost, an epistemic activity structured by epistemic goals. Thus, the accidental aesthetic payoff view acknowledges that paleoscience has an aesthetic side, but sees that as merely incidental to the core epistemic business of science. The fact that the practice of paleoscience has aesthetic *payoff* does not imply that it has aesthetic *goals*.

Even the accidental aesthetic payoff view implies that the paleosciences have an important aesthetic dimension. If epistemic practices have aesthetic payoff, and (as I argue later in Section 6) aesthetic practices have epistemic payoff, then there is considerable potential for positive feedback effects. Thus, the accidental aesthetic payoff view is compatible with what I earlier called the intertwining view. There are, however, two other potential problems with the accidental aesthetic payoff view.

First, there are areas of paleontological research where aesthetic goals are at times clearly in the driver's seat – or at least sharing the driver's seat with epistemic goals. For example, consider decisions involved in fossil collection (Currie 2017c). In places like Canada's Dinosaur Provincial Park, there is an abundance of dinosaur fossils. Many of these are just lying around on the surface, and would be fairly easy for researchers to collect. Decisions about what to collect and what to leave in (or on) the ground are complex, and involve at least three distinct sorts of considerations:

- *Epistemic considerations.* Is the specimen relevant to anyone's ongoing research project? Does it contain new information about the past?
- *Aesthetic considerations.* Is the specimen especially well-preserved, or beautiful, or complete? Would it be an exciting one to put on display?
- *Practical considerations.* What are the available resources for collecting and preparing the specimen? How much effort and expense would be involved in collecting it, storing it, and preparing it? And what will happen to the specimen if it's not collected?

Decisions about what to collect and what to ignore are, in a way, decisions about what to foreground and what to set aside as background (compare Wylie 2015, p. 41; Galison 1987). Fossil collection is, of course, data collection, and that makes it an epistemic practice. But it would be a mistake to construe fossil collection as an epistemic practice that just happens to have aesthetic payoff in the form of dramatic museum exhibits. Collection decisions are heavily (though not exclusively) guided by aesthetic norms and goals. Some of the reasons *not* to collect a *Hadrosaur* fossil, like the one pictured in Figure 1, are distinctively aesthetic. Museums have lots of them already, and this particular specimen is not especially exciting or thrilling. One might even think that a *Hadrosaur* fossil like that in Figure 1 has more aesthetic value when left *in situ*.

Second, recall the earlier distinction (from Section 1) between the "intertwining view" and the "blurring view." According to the latter, epistemic and aesthetic values blur together. I suspect that the blurring view is correct, and if so, it undermines the accidental aesthetic payoff view. In fossil collecting contexts, the completeness of a specimen is hugely important. Finding a *T. rex* tooth is exciting, but it's not the same thing as finding a skeleton that, like *Tyrannosaurus* Sue, is over 70% complete. Is completeness an epistemic value or an aesthetic one? It is pretty obviously both. Relatively complete specimens make for more exciting displays, but they also contain more information. So the norm, "Collect specimens that are relatively complete," is not readily classifiable as either epistemic or aesthetic. More generally, historical cognitivism pulls in the direction of the blurring view. For if aesthetic engagement with fossils and landscapes is always epistemically and cognitively inflected, then scientific investigation of the deep past starts to look like a mode of aesthetic engagement. Science is not an epistemic activity that just happens to have aesthetic payoff. Rather, science just *is* a way of engaging aesthetically with nature.

Traditionally, philosophers of science have tended to think of aesthetic value in the narrowly focused context of theory choice (McAllister 1989; O'Loughlin and McCallum 2019).[10] Things usually get framed in the following way. Suppose we have to choose between two rival theories, *T1* and *T2*. Suppose that those theories account for the data about equally well. But *T1* has some additional virtues. It is simpler, say, or more elegant, or more coherent. Some of these criteria of theory choice can look a lot like aesthetic qualities, and philosophers disagree about whether they carry evidential weight.[11] One prominent example of this is Lipton's (2004) notion of explanatory "loveliness," which he distinguishes from likeliness. According to Lipton, "loveliness" is a feature that can make one potential explanation better than others. Notice, however, that although this *looks* like a discussion of the role(s) of aesthetic values in science, it is actually a symptom of the one-sided focus on epistemology that I hope to push back against here. To start with, note that the focus here is on the aesthetic appreciation of *theories* or *explanations*, often with an emphasis on the beauty of mathematical structures (e.g. Engler 1990). There is something right about this. Part of the joy and excitement of theoretical work in science (and elsewhere) derives from the fact that a good theory can be a beautiful thing indeed. And creative theory construction and conceptual innovation have an aesthetic side, too. But if our goal is to better understand the aesthetic dimensions of scientific practice, this theory-centric debate barely scratches the surface.

Larry Laudan (2004) has claimed that "science is neither exclusively nor principally epistemic" (p. 15). Laudan's argument for this claim, however, focuses narrowly on the criteria of theory appraisal; he thinks that "many, and arguably most, of the historically important principles of theory appraisal used by scientists have been, though reasonable and appropriate in their own terms, utterly without epistemic rationale or foundation" (Laudan 2004, p. 16). This privileging of questions about theory appraisal still looks like an epistemology-driven approach to understanding science, since theory appraisal is basically a problem of deciding what to believe. Laudan and others may be right that aesthetic and other non-epistemic values play a central role in theory appraisal.

---

[10] Ivanova (2017) offers an overview of this traditional discussion of aesthetic value in science. Most of the literature she surveys is theory-centric, and there is little effort to engage with either environmental aesthetics or with historical science. However, see Catherine Elgin's (2002) fascinating analysis of the similarities between conceptual reconfiguration in art and science.

[11] For example, Van Fraassen (1980) famously argued that these so-called non-empirical theoretical virtues are merely pragmatic, and that they give us no reason for thinking that a theory is true or probably true. Many scientific realists, such as Schindler (2018), argue that these qualities are truth-conducive or truth-indicative.

But if we really want to understand the aesthetic dimensions of the practice of paleontology, we need to look beyond traditional questions about how to evaluate scientific theories.

The suggestion that scientific investigation has an aesthetic dimension could well raise some scientists' hackles. Many people think that aesthetic value is largely subjective; beauty is in the eye of the beholder, and all that. And because we lack any agreed-upon methods for adjudicating aesthetic disagreements, it can easily seem like aesthetic claims are mere expressions of subjective preferences or biases. In environmental contexts, for example, we might think that mountain scenery is lovely, whereas wetlands in the New England woods are mucky and gross. Or we might think that while the local chipmunks are cute, the bats in the neighborhood exude unpleasantness. Dinosaurs are dramatic and impressive; sponges, not so much. But these nature-directed aesthetic preferences might just reflect our own subjective biases, and those biases seem rationally indefensible – an issue to which I return in Section 9.[12] Science, however, is supposed to be objective. Of course, "objectivity" can have many meanings. But one of those meanings is that scientific claims are typically testable via procedures designed to reduce the impact of precisely these sorts of subjective preferences or biases. Looked at in this way, it's easy to see why scientists might take a dim view of anything to do with aesthetics. The very suggestion that scientific practice has an aesthetic dimension might seem like a threat to the rationality and objectivity of science.[13]

In reply to this worry about subjectivity, note that the whole point of Carlson's scientific cognitivism (and by extension, of the historical cognitivism I am defending here) is to show that aesthetic judgments have more objectivity than many people think. Those with empirical knowledge are in a better position, objectively speaking, to appreciate landscapes and fossils. This means that far from compromising scientific objectivity, aesthetic engagement with nature depends on it. My claim that the practice of historical science has an aesthetic dimension is thus not really about irrational, subjective biases infecting scientific work – although, as we'll see in Section 9, that can happen. The claim, rather, is that aesthetic engagement with landscapes and fossils is enriched by scientific knowledge of the past.

---

[12] This is a special problem for philosophers who, like Russow (1981) or Sober (1986), want to appeal to aesthetic values in generating arguments for environmental protection.

[13] There are parallel questions about social and political biases in science. Longino (1990) argues that contextual social and political value commitments need not undermine scientific objectivity. I think something similar is true of aesthetic commitments.

2.2 How Artistic Practices Contribute to Scientific Investigation

Paleoepistemology and paleoaesthetics are interdependent. The problem with much of the work that we philosophers have been doing is that we have tended, up to this point, to take paleoepistemology in isolation. Historical cognitivism shows how aesthetic engagement with fossils and landscapes depends on scientific knowledge of prehistory. But you might think that the relationship is one-sided. How does scientific knowledge depend on aesthetic engagement? What's the basis for the claim about *inter*dependence? Later, in Sections 4 and 6, I'll give some further examples of how aesthetic engagement can contribute to and motivate scientific investigation. But one example of paleontological work that shows how a certain sort of aesthetic engagement contributes to scientific knowledge is fossil preparation.

Caitlin Wylie (2009, 2015) has done extensive research on fossil preparators, the technicians – often working in museum labs – who take fossils collected from the field and "prepare" them for display and/or study. Fossil preparators are not necessarily scientists. Many do not have science degrees, and Wylie notes that paleontologists often do not include preparators as authors on their published papers. When you go to a natural history museum and see technicians on the other side of the glass, working carefully with drills and brushes to remove the matrix from fossils, those technicians are probably preparators, rather than paleontologists. Their work is essential, not just because they ready fossils for exhibition, but because they are producing data for paleontological research.

Fossil preparation has an epistemic dimension. Preparators have to make judgments about what is fossil material and what belongs to the surrounding rock – what to foreground, and what to set aside. Wylie (2015, p. 41) notes that this is very similar to the "laboratory judgment" that Peter Galison (1987) thinks experimentalists exercise when they try to sift the important experimental results from the background noise. Sometimes, preparators might get this wrong. It is likely that in earlier times, technicians may have removed dinosaur feather impressions, for the simple reason that no one thought of dinosaurs as the kinds of creatures that may have had feathers. So if there were feather impressions they may well have been removed in some cases along with the surrounding matrix. Fossil preparation is informed by background beliefs about the organisms in question, and about processes of preservation. And it's precisely because prepared fossils are the data of paleontology that preparation has such epistemic import. But as Wylie's work has shown, it also has an aesthetic dimension. Many fossil preparators have training in the arts, especially sculpture. And they often tend to describe their work in artistic terms. Much as a sculptor might try to visualize a statue inside a block of stone, and then chip away the surrounding stone to reveal the imagined form, a fossil preparator has to try to work out where the fossil specimen is within the matrix, and then reveal the fossil by a process of removal. Wylie also found that when preparators have to make tough decisions – for example, decisions about how exactly to glue a piece of fossil back together with another piece – they may be guided as much by aesthetic considerations as by anything else. She observes that "preparators believe that their work can change specimens' aesthetic categories, and in a sense they strive to improve specimens' beauty" (Wylie 2015, p. 39). In short, the practice of fossil preparation is in large measure an aesthetic practice. Because preparation is so time-consuming and resource-intensive, decisions also have to get made about which fossils to prepare, and those decisions, too, are aesthetic, made sometimes on the basis of judgments about which fossils would be most exciting to display.

In characterizing fossil preparation as an aesthetic practice, I mean to fold in at least five different features, all of which are implicit (and sometimes explicit) in Wylie's discussion. First, the practice is cognitively informed and mediated: Nothing that preparators do with fossils makes sense except against the background of beliefs about how those fossils are formed, about their ages, and so on. Second, the practice also involves aesthetic judgment and assessment: Preparators might decide to stop work on a specimen once it "looks good," or they might decide that a fossil looks better when glued together with a particular adhesive. But those aesthetic judgments are not free-floating, and interact with biological background knowledge in complex ways. Third, the practice involves a good deal of perceptual discernment and artistic skill. One must, for example, practice *seeing* the difference between fossil and matrix. Fourth, there are aesthetic goals in play, especially when preparing fossils for museum exhibits; the goal is to create prepared specimens that will impress visitors and draw their attention. Finally, like many artistic practices, fossil preparation is also an *embodied, material practice*, in which technicians use hand tools and interact materially with fossils. In this case, the material objects in the lab help shape the technicians' behavior in complex ways. For example, the type of rock involved can dictate which tools are appropriate, and also how long a project might take.[14] In all of these ways, fossil preparation is an aesthetic practice, and

---

[14] Alison Wylie (2002) is one of the few philosophers of science who has emphasized materiality in the context of historical science, though some philosophers and historians of science who focus more on experimental practice also emphasize materiality (e.g. Rheinberger 1997). Materiality theorists in archaeology, however, have long been sensitive to the ways in which material objects can shape human practices (for one interesting example, see Gell 1998).

one that contributes to the epistemic successes of paleontological research by making fossils available for study.

Wylie has shown, in the most vivid ways, that fossil preparation is a practice in which aesthetic and epistemic considerations collaborate. This is the case with other aspects of paleontological practice, too, such as the production of casts of fossils (both for exhibit and as a way of sharing data), and more recently, digitization and 3D printing of fossils (Cunningham, Rachman, and Lautenschlager 2014; Jones 2012; Rahman, Adcock, and Garwood 2012).[15] This aesthetic dimension of paleontology may be difficult to see, however – we philosophers have certainly been slow to recognize it – because, as Wylie also argues, some aspects of the social organization of paleontological research render the work of fossil preparators invisible. The practice of not giving preparators credit in scientific publications is one aspect of this. In what follows, I take Wylie's main insight about fossil preparation – namely, that its aesthetic and epistemic dimensions are inseparable – and argue that it extends much more widely. What's true of fossil preparation is true of a great deal of research in paleontology and the earth sciences.

To summarize, in this section I have sketched the bidirectional argument of the Element. On the one hand, historical cognitivism shows how the epistemic dimension of the paleosciences contributes to aesthetic engagement with fossils and landscapes. On the other hand, Caitlin Wylie's work shows how aesthetic practices such as fossil preparation make an indispensable contribution to scientific inquiry. This suggests that the aesthetic and epistemic dimensions of paleoscience render mutual support. But that, in turn, suggests that it is a mistake to try to work out the epistemology of historical science without also exploring palaeoaesthetics.

3 Historical Cognitivism in Aesthetics

In Section 2, I began making the case that paleoaesthetics and paleoepistemology are interdependent. This is a two-way argument; on the one hand, knowledge of the deep history of fossils and landscapes enhances our aesthetic engagement with them. On the other, aesthetic values and practices also contribute to the epistemic successes of paleoscience. In this section, I develop the first line of

---

[15] Tamborini (2019) argues that we should think of paleontology as a technoscience. That seems right, though it's also worth exploring how some of the relevant technologies, such as 3D printing, also figure in artistic practices.

argument in more detail. I'll spell out historical cognitivism more carefully, introduce a new argument for it, and address some potential objections.

In developing his account of scientific cognitivism, Allen Carlson draws an analogy between art criticism and natural science:

> If to aesthetically appreciate art we must have knowledge of artistic traditions and styles within those traditions, to aesthetically appreciate nature we must have knowledge of the different environments of nature and of the systems and elements within those environments. In the way in which the art critic and the art historian are well equipped to aesthetically appreciate art, the naturalist and the ecologist are well equipped to aesthetically appreciate nature (Carlson 1977, p. 273).

On one strong reading of this view, scientific knowledge is a necessary condition for the proper aesthetic appreciation of nature. This strong reading might, however, be too strong. Emily Brady, for example, argues that "cognitive models run into problems when they make science a necessary framework and the only correct one" (Brady 2003, p. 99). One problem is just that there would seem to be counterexamples. Surely it is possible to have certain kinds of valuable aesthetic experiences without having the requisite scientific knowledge. Imagine if you were raised (as I was) in a place with flat topography and almost no interesting exposed geology, transported suddenly to the badlands, where different rock strata are plainly visible. You might have no knowledge of geological history, no understanding of the principle of superposition – i.e., the principle that lower strata are older – and no understanding of the processes by which different strata are formed. But you might still gaze in wonder at the visual scene, and even wander around the landscape, getting a feel for the place and carefully studying the features of the exposed rock. You might sketch or photograph what you see. Let's call this *naïve engagement*. It would be crazy for the cognitivist to deny that naïve engagement with fossils or with landscapes has any value, just as it would be crazy to deny that naïve engagement with music or art has value. Brady is right; if the suggestion is that (a certain amount of) scientific knowledge is a necessary condition for proper aesthetic appreciation, then cognitivism seems rather implausible. After all, as I will show later, in Section 4, naïve engagement is often the starting point for empirical investigation. And even though naïve aesthetic engagement does not require knowledge, it does nevertheless involve thought. One reason why I am using the term "engagement" here, even though some theorists who talk about aesthetic "engagement" are non-cognitivists (e.g. Berleant 1995), is that the term captures the way in which even someone without much empirical knowledge is active. Even if you are a naïve engager with the landscape, you are thinking about what you are seeing and touching.

A more promising way to develop Carlson's basic idea is to contrast naïve with informed engagement:

> *Historical cognitivism in aesthetics*: Knowledge of living things and natural systems – including knowledge of the history of those things – deepens and enhances our aesthetic engagement with those things, relative to various kinds of naïve engagement.

There is nothing here about scientific knowledge being a necessary condition for proper aesthetic appreciation. Instead, the claim is that knowledge, especially knowledge of the past, makes us better able to appreciate nature's aesthetic qualities.[16] This is compatible with acknowledging the value of various kinds of naïve aesthetic engagement, and lends itself to a more pluralistic view, along the lines of Hettinger (2008). Historical cognitivism has one other significant limitation, in that it only focuses on one kind of knowledge. A variety of other kinds of knowledge, including tacit knowledge-how (as contrasted with knowledge-that) come into play in scientific research, and could also enhance aesthetic engagement. With this in mind, we should not construe historical cognitivism as a comprehensive account of how knowledge contributes to aesthetic engagement.

Historical cognitivism in aesthetics, as formulated here, is somewhat weaker than Carlson's view. However, one might still wonder if it is too demanding. One might think that what really deepens aesthetic engagement is not knowledge of history, but something more like investigative effort or genuine epistemic engagement.[17] The question is what takes us further than naïve engagement. So far, I've argued that knowledge of history, or epistemic success, takes us further. But what if *trying* to learn about the past is more important than actually succeeding? On such a view, sincere, effortful investigators could also have richer aesthetic appreciation of things, even if they don't (yet) know the history of those things, and even if they have some badly mistaken beliefs. Still, there are reasons for thinking that it is actually knowledge of history that really matters. It is important to leave open the possibility that historical knowledge— perhaps acquired via a museum exhibit or imparted by a scientific guide—can enrich the aesthetic experience of non-specialists and non-scientists who are not

---

[16] This is close to Yuriko Saito's (2008) view, although she also goes further than Carlson in adding that science helps us to "appreciate nature on its own terms," which she argues is an ethically better form of appreciation. See especially Saito (2008, p. 156). Glenn Parsons (2008b, p. 305) also opts for a weaker version of historical cognitivism. His version is that "scientific knowledge is necessary for the correct appreciation of *some* of the aesthetic qualities" of any natural object. Parsons's version of the view strikes me as too weak, because it doesn't imply that increasing scientific knowledge further enhances aesthetic appreciation.

[17] I am grateful to Adrian Currie for making this interesting suggestion.

themselves actively investigating things. Plausibly, there are also cases in which sincere, effortful investigation misfires in ways that make it worse than naïve aesthetic engagement. Consider as an example an effortful investigator who sincerely believes that aliens visited Earth a long time ago and imparted architectural and technological know-how to ancient people. This amateur investigator is sincerely committed to the ancient aliens hypothesis, and visits sites – Chaco Canyon, say – looking for evidence that professional archaeologists may have missed. But the believer in ancient aliens also has very little knowledge of the history of those sites. This person's aesthetic engagement with those special places is limited by ignorance and corrupted by massive misinterpretation. Far better to be a naïve appreciator of special places than someone with no more knowledge than a naïve appreciator, but with an outlandish conspiracy theory.

The case in favor of historical cognitivism need not rest entirely on thought experiments involving qualitatively identical items with different histories. For there are real paleontological cases in which we've had to rethink the histories of things in ways that make an obvious aesthetic difference.

In the summer of 2014, the auction house I.M. Chait put an unusual item up for bid (Switek 2014). The object is about 6 million years old. It sold for $10,370, a price that included a pretty hefty commission, to a collector who clearly thought that it has some aesthetic value. According to the auction house, the item was a coprolite, or fossilized feces:

> This truly spectacular specimen is possibly the longest example of coprolite ever to be offered at auction. It boasts a *wonderfully even, pale brown-yellow coloring and terrifically detailed texture* to the heavily botryoidal surface across the whole of its immense length. The passer of this remarkable object is unknown, but it is nonetheless a highly evocative specimen of unprecedented size, presented in four sections, each with a heavy black marble custom base, an *eye-watering 40 inches in length* overall.[18]

This description emphasizes the item's aesthetic qualities, and does so on the assumption that it's a coprolite. For example, if it weren't a coprolite, it's not clear that it would make sense to play up the "immense length" or the "unprecedented size." Our aesthetic appreciation of the object depends on what we believe it to be. And our beliefs can turn out to be false.

As it happens, this object is probably not a coprolite at all. It was collected from the Wilkes Formation in southern Washington state, which is rich in

---

[18] Emphasis added. The auction house catalog is no longer publicly accessible, but the text was quoted in news reports at the time, such as Sharwood (2014), who predictably misattributes the coprolite to a dinosaur.

"coprolites." You can find many for sale online. But scientists who've looked into this are not sure that they are coprolites at all. In a paper published way back in 1993, Patrick Spencer proposed a completely different explanation for the occurrence of the would-be coprolites (Spencer 1993; see also Mustoe 2001). Imagine a rotting hollow log. If the log were buried with organic-rich silt and clay, physical forces could squeeze the sediment through the knot-holes, creating something that looks a lot like a coprolite. In a more recent interview with Brian Switek, Spencer reports that he'd tried cross-sectioning a number of Wilkes Formation "coprolites" to see if there was any residue from the animal's last meal, but "never found a dang thing in there." Buyer beware!

Now, someone paid a great deal of money for this object, presumably in part because of its unique aesthetic qualities. One purchases this sort of thing (I guess) to display on the mantelpiece and show off to dinner party guests. Or perhaps to donate to a museum where others can appreciate it. It is, as it were, a "collectors' item," much like an artwork or a rare autograph. But the empirical evidence simply does not support the view that the object in question is really a coprolite. The buyer very likely has a false belief about the causal history of the item. The buyer's aesthetic engagement with the object – the seeing it *as* a coprolite, and describing it on that assumption, while marveling at its length – is going awry and getting things significantly wrong. The buyer is a little bit like the creationist who goes to the Grand Canyon and marvels about how Noah's flood could have carved out such a quantity of rock. There is some aesthetic engagement taking place, but it is misfiring badly. The suggestion here is that false beliefs about the past undermine aesthetic engagement in a significant way. I think the auction house "coprolite" case strongly supports that suggestion. But if the suggestion is right, then historical cognitivism must be correct. For historical cognitivism is just the view that aesthetic engagement is (partly) a matter of beliefs about the past.

Notice how well this case mirrors Robert Elliot's thought experiment involving the fake Vermeer. The argument for historical cognitivism is not merely thought experimental. This case shows that the value of a fossil does depend in part on its causal history. It also shows how historical cognitivism introduces some objectivity into aesthetic judgment. False beliefs about an item's history can lead to various kinds of misappreciation, and to the over- or undervaluing of its aesthetic qualities. The basic argument in favor of historical cognitivism is just that it is the only view that can make sense of such cases, where revising our beliefs about an item's history completely changes the character of our aesthetic engagement with that item. Aesthetic judgments about things are sensitive to beliefs about history.[19]

One other possible line of argument for historical cognitivism takes its inspiration from the experience of playing or listening to music. We might call this the *musical analogy*. Imagine being transported into a concert hall, so that all you hear is the closing bar of an unfamiliar symphony. Of course, that naïve experience of the last chord might be rich and full in its own right. But surely your appreciation of the last chord would be enhanced – and you would become a better judge of what you're hearing – if you understood how it fits into the larger piece of music. The reason for this is that the chord is just part of the larger, temporally extended piece of music, and the larger piece of music is the proper target of aesthetic engagement. Obviously, there are salient differences between a musical composition and geological processes, but there is also a relevant similarity. Fossils and rock outcrops are a lot like the last chord of the symphony. One can appreciate them and engage with them naively, having no idea of what came before, and there may be some value in that. But perhaps, as in music, the proper target of aesthetic engagement in historical science is the larger, temporally extended process. If we think of unfolding historical processes, rather than currently existing items, as the bearers of aesthetic value, then it's easy to see how historical knowledge matters to aesthetic engagement. The purchaser of the would-be coprolite is like someone who heard the last bar of a symphony, but completely misunderstood what the larger piece of music must have been like.

In the remainder of this section, I'll consider some potential objections against historical cognitivism.[20]

3.1 Elitism

One might reasonably worry that historical cognitivism commits us to a form of elitism. Emily Brady, for example, complains that Carlson's scientific cognitivism "threatens to leave out non-expert judgments" (Brady 2003, p. 99). Cognitivism in effect privileges the aesthetic judgments of those few who have scientific knowledge. Even the weaker version of cognitivism that I am defending here – the version that stops short of making knowledge of history a necessary condition for proper aesthetic engagement – still insists that those with knowledge are better positioned to appreciate landscapes, fossils, and other things in nature, that their engagement with nature is richer.

---

[19] This argument for aesthetic cognitivism parallels arguments for cognitivist theories of the emotions. Cognitivists about the emotions (e.g. Nussbaum 2003) point out that emotions such as anger are sensitive to changes in the evidence.

[20] For further discussion of these objections, see Parsons (2008a, pp. 59–65).

The charge of elitism would only stick, however, if scientific knowledge were some sort of rivalrous good, such that my having more of it must mean that others have less. But scientific knowledge is readily shareable with as many as possible. Once you see this, the initial worry about elitism actually points to a surprising virtue of historical cognitivism. Cognitivism helps explain why science education and public outreach are so important. As people learn more about evolutionary and geological history, they develop a fuller sense of place, and a better appreciation of the plants and animals around them. Historical cognitivism means that there are additional aesthetic reasons for disseminating knowledge of geobiological history as widely as possible – and that seems like the very opposite of an elitist view.

Another version of the worry about elitism is that it privileges Western science over indigenous perspectives. This worry may itself have an elitist flavor though, since it seems to assume that indigenous traditions do not transmit as much knowledge of natural history. Historical cognitivism makes no such assumption. Indigenous traditions include a lot of empirical knowledge of regional environments, and that alone makes for richer aesthetic engagement. For example, when European settlers moved into the Connecticut River Valley in the 1600s and 1700s, they were almost entirely clueless about the geological and ecological history of the region. But Native American communities who'd lived there for thousands of years had oral traditions that included extensive knowledge of landscape history. For example, oral traditions included "earth-shaper" stories in which giant characters – such as a giant beaver – modified the landscape. Bruchac (2005) discusses one such story that looks a lot like an account of the draining of glacial Lake Hitchcock – a fascinating convergence of indigenous oral tradition and geological insight. Also, Native American narratives about the landscape altering activities of giant beavers have to be seen in the light of paleontological findings that there really were giant, 7-foot-long beavers altering the landscape. If anything, historical cognitivism may help explain why these indigenous intellectual traditions are so important, and why they deserve a lot of deference.

Another response to this second version of the worry about elitism is that historical cognitivism highlights some under-appreciated similarities between indigenous and scientific traditions. Yuriko Saito observes that "both scientific explanation and folk narratives are attempts at helping nature tell its story to us ... " (Saito 2008, p. 163). One might wonder about the sharpness of the distinction between scientific explanation and folk narrative, but Saito makes an important point. Both use knowledge of the histories of places to deepen our connections with those places. There may of course be cases where indigenous traditions disagree with the findings of natural science. But in those cases, historical cognitivism does not say that we should just assume that the natural science is correct. Historical cognitivism only says that knowledge of the past enhances aesthetic engagement. Notwithstanding Carlson's use of the term "scientific cognitivism," historical cognitivism need not say *where* to find that knowledge of the past. Far from being elitist, a historical cognitivist view in aesthetics helps to show why indigenous intellectual traditions should be taken seriously.

3.2 Positive Aesthetics

Some philosophers have thought that scientific cognitivism in aesthetics has unpalatable logical implications. Some think that it implies a thesis known as *positive aesthetics* – or the thesis that everything in nature has positive aesthetic value. Consider a case of something that initially appears yucky or gross – such as termites or deer ticks. The thought is that as you learn more about these organisms and come to understand their biology, they will no longer seem so yucky or unpleasant. You might come to think that they have lots of positive qualities. Some philosophers, like Carlson (2008), have embraced this alleged consequence, seeing it as a virtue of the view. Parsons (2008b), too, is happy to endorse positive aesthetics. If what we want, from a policy perspective, is an argument for protecting natural systems and biological diversity on aesthetic grounds, then positive aesthetics seems like just what the policymaker ordered. Some argue, however, that positive aesthetics is just implausible on the face of it, and that some things in nature certainly are ugly or gross (Brady 2011).

Even if historical cognitivism did imply positive aesthetics, I'm not at all sure that would be a problem. However, it's also not clear to me at all that historical cognitivism has that implication. We've already seen cases, like that of the misidentified "coprolite" or the duplicate Vermeer, where learning about the history of some item makes us realize that we'd overvalued it aesthetically. Sometimes, learning about the history of something can reveal negative—or even horrifying – aesthetic qualities. For example, in the film, *The Red Violin*, we learn something macabre about how the violin was made that completely alters our assessment of the instrument's "beauty" – and not in a good way. Sometimes positive aesthetic qualities are inextricably bound up with negative ones, and learning about the history just reveals those fascinating connections.

The deeper problem here may be the background idea that aesthetic theory is all about beauty, or that beauty is somehow *the* aesthetic value. Someone in the grip of this idea might see historical cognitivism as committed to the view that everything is beautiful in virtue of its history. I think we should reject some of the background assumptions that help to generate the worry about positive aesthetics in the first place. Aesthetic engagement involves much more than just appreciating something's beauty. Knowledge of history deepens aesthetic engagement, but that can mean appreciating positive aesthetic qualities other than beauty – think about how knowing where your food came from can make it more delicious. Knowing where your food came from can also make that food disgusting. The knowledge deepens your aesthetic engagement with your meal in ways that need not be closely linked to notions of beauty.

3.3 Science as a Source of Disenchantment

There is a strand of thinking in Western culture, associated with the romantic movement, according to which scientific knowledge leads to disenchantment. Stan Godlovitch complains that "science demystifies nature by categorizing, quantifying, and patterning it" (Godlovitch 2008, p. 142). As scientists learn more about causes and effects, mystery recedes, and nature becomes denuded of meaning. In "The Tables Turned" (1798), William Wordsworth expresses this concern:

> Sweet is the lore which Nature brings;
> Our meddling intellect
> Mis-shapes the beauteous forms of things:—
> We murder to dissect.
>
> Enough of Science and of Art;
> Close up those barren leaves;
> Come forth, and bring with you a heart
> That watches and receives.

The concern here is that the intellect actually gets in the way of our appreciation of "the beauteous forms of things." One version of this concern is that science distracts us from the sensuous complexity of nature. If you are thinking too hard about how glaciation shaped the New England landscape, you'll fail to notice the beauty of the moss-covered boulders strewn around the woods. Another, slightly different way in which science might get in the way of aesthetic appreciation of nature is by removing the mystery from things. If you know all about dinosaurs, for example, it is certainly exciting to see a hadrosaur bone eroding out of the ground. But the experience might be more amazing if you had little or no idea what you were looking at—if you were gazing in wonder at something strange. The larger worry is that scientifically informed aesthetic engagement is not necessarily any better than – and might be worse than – naïve engagement.

One problem with the romantic objection is that if it were a good argument, it would prove too much. If, in general, empirical knowledge gets in the way of aesthetic engagement with the natural world, then that would be one reason to eschew scientific knowledge. Notice how this reply dovetails with my earlier reply to the elitism objection. There I pointed out that historical cognitivism has the virtue of explaining why it is especially important to share scientific knowledge as widely as possible – for example, in natural history museums. The romantic objection would seem to have pretty radical implications on this front, for it implies that we have aesthetic reasons not to share scientific knowledge. A proponent of the romantic objection would seem to be saying that museum-goers would have better aesthetic experiences of fossils if they knew nothing about what they were seeing.

A second problem with the romantic objection is that it rests on a mistaken view of science's relationship to mystery and enchantment. According to that (mistaken) view, scientific investigation is a little bit like exploring and mapping a cave. At first, it's a mystery how deep and how large the cave might be. But as the spelunkers proceed to map things carefully, the mystery dissipates, and the cavescape is rendered more familiar. On this sort of model, mystery and wonder decline inexorably as science makes progress. However, this is not really how science works. Often, as we learn more about the world, or about the past, new mysteries open up, and things seem even stranger and more wondrous than before.[21] If the spelunkers feel some sense of loss at the thought that formerly mysterious regions have been mapped, this is quickly superseded by new forms of puzzlement – for example, over when and how certain chambers were formed, or about the evolutionary histories of the things living there.

The late Cretaceous clam bed from Section 2 is an excellent illustration of what's wrong with the romantic objection. For someone with some knowledge of the landscape and its deep geological history, there is a fascinating mystery about evolutionary stasis. Why have freshwater bivalves changed so little, morphologically, over tens of millions of years, even while everything else around them has changed so dramatically? This is the sort of mystery that a disciple of Wordsworth might never wonder about, because it's one that only shows up against the background of a lot of scientific knowledge. If anything, having some knowledge of prehistory makes the world around us even more enchanting.

To drive this last point home, consider the work of another poet, Emily Dickinson:

> Volcanoes be in Sicily
> And South America
> I judge from my Geography

---

[21] This is also a central theme of Dawkins (1998).

Volcanoes nearer here
A Lava step at any time
Am I inclined to climb
A Crater I may contemplate
Vesuvius at Home

When Dickinson writes of "volcanoes nearer here," she was very likely referring to the landscape near her home in Amherst, Massachusetts. Some scholars think that she probably attended lectures by Edward Hitchcock at Amherst College, and it is also likely that she used Hitchcock's geology textbook in school (Uno 1998). Hitchcock, at the time, was probably the leading paleontologist in North America. Today, he is best known for his collecting of dinosaur footprints from around the Connecticut River valley – though he didn't refer to the trackmakers as "dinosaurs." Hitchcock (1878/1974) called them "Brontozoa" and supposed (quite presciently, it turns out) that the tracks had been made by giant prehistoric birds. Hitchcock also knew well that the landscape of the Connecticut Valley was volcanic. During the late Triassic and early Jurassic, around the time that the dinosaurs were leaving those footprints, what is now the Connecticut valley was a volcanically active rift valley, much like the east African rift valley today. There is evidence of this volcanism in my own town, where a strange ridge known as the Higganum basalt dike runs for many miles, southwest-to-northeast (Philpotts and Asher 1994). It was formed around 200 million years ago by laterally flowing lava. And just south of Amherst, Massachusetts, there is a volcanic ridge – bisected today by Interstate 91 – that Dickinson could easily have climbed, "a lava step at any time."

Dickinson's writing here serves as a kind of poetic rebuttal of the romantic objection (and see Heringman 2004, who makes a similar point about British romantic poetry's debt to geology). She is expressing a sense of place in western Massachusetts that's rooted in geo-historical knowledge that she could have gotten from Edward Hitchcock. Knowing that the landscape around you was formed by prehistoric volcanism changes how you situate yourself in that landscape. It's implausible that scientific knowledge, in this case, is contributing to disenchantment, or getting in the way of aesthetic engagement with place.

In summary, the main argument for historical cognitivism is that it is the only view that can make sense of our aesthetic responses in cases like that of the 40-inch "coprolite." I have also sketched initial rejoinders to the worries about elitism, positive aesthetics, and disenchantment. I've defended a version of historical cognitivism that's weaker than Carlson's, and allows for the importance and value of certain kinds of naïve aesthetic engagement while insisting that other things being equal, scientific knowledge of the past tends to deepen and enrich aesthetic engagement.

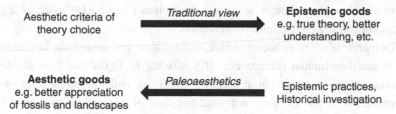

Figure 4 Contrasting paleoaesthetics with the traditional privileging of epistemic goods.

Once you see how historical scientific knowledge can enhance aesthetic appreciation, that has profound implications for how we understand the practice of science. In Section 2 above, I noted how philosophers of science have traditionally approached questions about aesthetic value in the context of thinking about criteria for theory choice. This way of framing things effectively subordinates aesthetic values to epistemological concerns. Aesthetic values turn out to be interesting, on this traditional view, only insofar as they carry evidential weight. The view I am advocating here turns this traditional approach upside down. If historical cognitivism is true, then scientific knowledge will turn out to have *instrumental* value insofar as it contributes to better aesthetic engagement with fossils and places. Knowledge is surely valuable for its own sake, too; the point is just that if we want to, we can easily subordinate epistemic concerns to aesthetic ones.

4 Aesthetic-Epistemic Feedback Effects

I have been arguing that historical knowledge deepens and enriches aesthetic engagement. Historical cognitivism suggests that scientific investigation of the past can contribute to aesthetic engagement with fossils and landscapes. However, the relationship between the epistemic and the aesthetic dimensions of historical science is bidirectional. Not only does historical research make for richer aesthetic engagement, but aesthetic engagement can also deepen and enhance scientific investigation. Aesthetic engagement and historical investigation are mutually facilitating. Another idea, borrowed from environmental philosophy, can help illuminate how this works.

Bryan Norton (1987) and Sahotra Sarkar (2005) both argue that transformative value helps to provide a rationale for protecting biological diversity. However, as they acknowledge, many things aside from endangered species can have transformative value. Indeed, both fossils and landscapes can have this kind of value. First, though, let's consider Norton's and Sarkar's line of argument concerning biodiversity conservation.

Demand value, or economic value, is the value that something has insofar as it satisfies human preferences. If you want to figure out how much it makes sense to invest in protecting (say) an endangered species, one approach is to try to figure out what the cost of the extinction would be. This is equivalent to asking what demand value the species has. And species do often have considerable demand value. For example, quinine is an important drug that was historically used to treat malaria – a drug that has saved countless lives. It derives from alkaloids found in the bark of the cinchona tree, which is native to South America. Historically, cinchona trees have therefore had considerable demand value. But it's not clear that in other cases, demand value provides such a compelling rationale for conservation. For there are many species whose economic value is far from obvious. Some environmental philosophers, worried that appeals to the economic value of species will not justify big investments in conserving biodiversity, take a non-anthropocentric turn. They argue that species (or more abstractly, biodiversity) have intrinsic value, or even that we have duties toward endangered species (Rolston 1989). Norton and Sarkar argue that one need not go so far. They defend what Sarkar refers to as a "tempered anthropocentrism." Transformative value, according to them, is an important kind of anthropocentric value that cannot be reduced to economic or demand value. That's because demand value is always relative to the preferences people happen to have, whereas transformative value is the value something has, insofar as it has the power to induce revisions of our preferences.

Although neither Norton nor Sarkar explicitly treats transformative value as a kind of aesthetic value, they tellingly use examples of aesthetic experience to illustrate what they have in mind. Sarkar, for instance, imagines the following sort of case. Suppose that you just have no desire to hear classical music. You've never made any effort to listen to classical music, aside from the unavoidable ambient snippets. The local symphony orchestra is performing this weekend, but the music has no demand value for you. If an economist asked you how much you would be willing to pay for tickets, the answer would be zero. If you do love classical music and find the example tough to relate to, then imagine the same case with some other less familiar art form: bluegrass, or modern dance, or whatever. Now a friend happens to have an extra ticket to the concert, and invites you to come along. Thinking that sitting through boring music is a fair price to pay for a pleasant evening with friends, you go along, and you hear (let's say)

Max Bruch's violin concerto in G minor. Your mind is blown. Something about the work speaks to you, and you find yourself getting lost in the music. Afterwards, you want to learn more, and you want to listen to more classical music: Who was Bruch (whom you had never heard of)? When and where did he live? Who influenced him? What else did he compose? Your preferences are transformed, and in the future you will twist your friends' arms to get them to accompany you to hear classical music. As your experience broadens and you learn more about music, the very nature of your experience changes over time. The performance, Sarkar and Norton might say, has transformative value.

If the goal is to generate a new line of argument for protecting biological diversity, this appeal to transformative value almost immediately runs into insuperable problems. Sarkar considers one of these, which he calls the *directionality problem*: What if the concert experience reinforces your aversion to classical music? Just how something transforms someone's preferences is a matter of contingent psychological fact, and there is no guaranteeing that someone's preferences will move in the direction we like. There is a deeper worry about circularity here too: How do we even know which direction of transformation would be good? Why would it be better if the concert leaves me loving classical music, than if it were to transform my preferences the other way? In order to answer that question, it won't help to appeal to the transformative value of the concert. We'd have to tell some other story about why classical music is so great. But then it's the other story, rather than the appeal to transformative value, that's really doing the philosophical heavy lifting. Often, it's experiences of bad things that have the most lasting transformative impact. If you witness human suffering in the aftermath of a natural disaster, that might lead you to pursue a career in medicine, or to become an aid worker. But it sounds perverse to say that the suffering has transformative value.

L.A. Paul's (2015) recent work might also help to highlight some problems with the Norton/Sarkar account, though from a slightly different direction. Paul is concerned with cases where we have to make major life decisions that could end up transforming our values and preferences. The basic issue is that it's not clear how our values and preferences are supposed to guide our rational decision-making, when a major life decision (such as whether to have children, or whether to change careers) will make a difference to what values and preferences we end up having.

As a gambit in environmental ethics, the appeal to transformative value is not too promising. Nevertheless, Norton's and Sarkar's account of transformative value does get the phenomenology right, even if it fails to provide a very robust argument for protecting biodiversity. Norton and Sarkar succeed in capturing one important way in which aesthetic engagement and investigation might mutually suffuse one another. On the one hand, your experience of the Bruch concert makes you want to learn more about classical music, and to experience more. But as you learn more about music, that deepens your aesthetic appreciation and further enriches your subsequent experiences of it. The insight at the heart of Norton and Sarkar's account is that there can be positive feedback loops between aesthetic engagement and epistemic investigation (Figure 6). Aesthetic experience motivates investigation, and investigation deepens and enriches aesthetic experience. Of course, the second half of the feedback effect – the part about scientific knowledge enriching aesthetic experience – presupposes scientific cognitivism of the sort I defended in Section 3. If my analysis is correct, Norton and Sarkar have captured an important aspect of human experience, but they get the emphasis a bit wrong when they talk about "transformative value" as something that resides in species or in biodiversity or in musical performances. It's more helpful to treat their story about transformative value as an account of a special kind of interactive process.

Using their terminology, however, it is easy to see how fossils and landscapes might have transformative value. In the simplest sort of case, just seeing a special fossil on display in a museum can have an impact on you, transforming preferences, provoking questions, and arousing curiosity. The process of learning more about what you've seen can then change your experience of those very same fossils.

One potential worry about the argument of this section is that there is a crucial distinction between: (a) *motivating* scientific investigation; and (b) *structuring* or *shaping* that investigation. The Norton/Sarkar account of transformative value shows how aesthetic goals and interests can motivate inquiry. But the claim that aesthetic goals sometimes shape the direction of inquiry, and structure the practice of science, is a more ambitious one. In the Section 5, I set to work making this more ambitious claim plausible

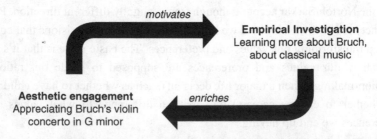

Figure 5 An epistemic-aesthetic feedback effect

5 Functional Morphology as Aesthetic Engagement

How might aesthetic goals help structure paleontological investigation? In order to see this more clearly, it will help to zero in on one particular kind of paleontological research. Functional morphologists seek to infer the functions of initially puzzling fossilized structures. For example, what were those bony plates on the back of *Stegosaurus* really for? What was their function? Was it thermoregulation? Defense? Display? Something else entirely? Interest in function often shapes paleontological investigation (Turner 2000). There is also a long tradition in aesthetics that sees a close connection between beauty and functionality (Parsons and Carlson 2008). Indeed, this linkage shows up frequently in biology – for example, in Darwin's famous reference to "endless forms most beautiful" at the end of the *Origin of Species*. More recently, George McGhee (2011) has given his book on theoretical morphology the subtitle: "Limited Forms Most Beautiful." Owing in part to this connection between functionality and beauty, the practice of functional morphology is a form of aesthetic engagement, structured by aesthetic goals.

In their book, *Functional Beauty*, Parsons and Carlson (2008) trace a strand of thinking in the history of Western aesthetics that sees beauty as having to do with fitness-for-function. For my part, I do not find the narrow focus on beauty – or even "aesthetic value" – as a quality of objects, to be too helpful. We should allow that there are lots of different positive aesthetic qualities that we might appreciate in objects. And appreciation of those positive aesthetic qualities is just one aspect of aesthetic engagement. Aesthetic engagement is a cognitively involved, embodied, two-way interaction in which objects also do things to us. Still, with these qualifications, Parsons and Carlson do make an important point about how knowledge of function bears on our aesthetic assessments of ordinary artifacts. For example, in snowy northern regions, people sometimes equip their bicycles with wide snow tires. If you did not know the function of the wide tires, you might find them somewhat comical. Knowing the function of the tires should change our aesthetic assessment of them; rather than looking comical or silly, they might evoke thoughts of outdoor winter activity. For another example, think about how you might assess a bicycle with a flat tire, its unfitness for function makes it look a little sad and forlorn. These mundane examples show that our aesthetic judgments are informed by knowledge of function.

In paleontological contexts, scientists often do not know the biological functions of the fossilized structures they have available for study. The argument of Section 4 above was that an encounter with an especially exquisite fossil can motivate paleontological research in much the same way that experiencing an amazing concert can motivate us to learn more about musical history and theory. We can now go a little further than that. Why is it that one of the first questions that paleontologists ask about a fossil is "what (if anything) was it *for*?" Philosophers of science have tended to assume that biologists are interested in function primarily because they care about functional explanation. For example, "*Stegosaurus* has those bony plates because the plates facilitate thermoregulation." However, Parsons and Carlson's line of argument points to an equally deep aesthetic interest in learning the functions of things. Functional hypotheses help guide and direct our aesthetic engagement with fossils. We could even think of a functional hypothesis as a proposal for engaging aesthetically with an object in a certain way. On this way of looking at things, functional morphology is a kind of aesthetic exploration. In order to make this a bit clearer, I'll develop a paleontological example in more detail.

Ammonoids were prehistoric cephalopods, with beaks and tentacles like today's squids, to which they are closely related.[22] But they also had coiled shells, as shown in Figure 5, and they would have looked quite a bit like their other living evolutionary relatives, the chambered nautiluses. The term "ammonoid" derives from the Egyptian god Ammon, who had horns like a ram. The soft parts of ammonoids rarely fossilized, but we have loads of ammonoid fossil shells. Their sheer abundance and variety – and the relative completeness of the ammonoid fossil record – is probably what the German paleontologist Adolf Seilacher had in mind when he wrote that "ammonoids are for paleontologists what *Drosophila* is in genetics" (Seilacher 1988, p. 67). Their long evolutionary run of 290 million years, give or take, together with their diversity and abundance, provides us with a good look at larger patterns of evolutionary history. Somehow, the ammonoids survived several mass extinction events, including the horrific "great dying" at the end of the Permian period, around 251 million years ago. The asteroid that probably finished off the dinosaurs around 66 million years ago, perhaps in conjunction with other upheavals at the end of the Cretaceous period, finally did them in.

Ammonoid shells were originally made of aragonite (a form of calcium carbonate), though the original material rarely preserves, and most ammonite fossils are just rock – usually sedimentary casts or molds. But sometimes the shelly material preserves. Today, in western Canada, there is a small ammolite extraction industry. Ammolite is an iridescent gemstone consisting largely of aragonite from Cretaceous ammonoids that once swam around the interior

---

[22] Although I use the term "ammonoid" here, which is the technical name for the taxon, many people use "ammonoid" and "ammonite" interchangeably. Some specialists reserve the term "ammonite" for the *Ammonitina*, a subgroup of the ammonoids. See Monks and Palmer (2002, pp. 107–8).

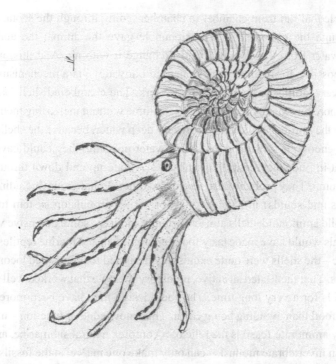

Figure 6 Typical Mesozoic ammonoid. Artwork by the author.

seaway that bisected North America. People mine the remains of ammonoid shells because they find that the material makes for beautiful jewelry.

People collected and admired ordinary ammonoid fossils long before anyone knew quite what they were. In parts of Europe, stories circulated about how the strange fossils were actually petrified snakes – snakestones – rocky relics of past extermination campaigns waged by this or that local saint. This is yet another case that illustrates the basic idea of historical cognitivism; what you think about the genesis of an item affects your aesthetic engagement with it. How you see an ammonite fossil – for example, whether you see it as a snake – depends on how you think it came about.

Intuitively, you might think that the main function of the shell was to offer protection from predators. Yet that is only one part of the story. Ammonoid shells had chambers, much like the shells of a chambered nautilus. The animal lived inside the last chamber. And like nautiluses, ammonoids built their shells chamber by chamber as they developed. Each chamber was separated from the next by a thin wall called the septum.[23] Ammonoids also had a long organ called

---

[23] Some of the technical jargon of ammonoid shell morphology can be daunting, but Monks and Palmer (2002, ch. 2) offer a helpful overview.

a siphuncle that ran from chamber to chamber, going through the septa, all the way through the coiled shell. The siphuncle gave the animal the ability to remove water from the chambers and exchange it with air. And this, in turn, made it possible for them to achieve neutral buoyancy via a mechanism that's actually very similar to how submarines work. The chambered shell made the animals more buoyant by increasing their volume without increasing their mass. Although they probably did not live in very deep water, because the shells were not thick enough to withstand that much water pressure, they could have used their built-in buoyancy control systems to navigate up and down through the water column. They probably also resembled other cephalopods (including both nautiluses and squids) in having some sort of jet propulsion system to boot. Thus, while ammonoid shells almost certainly did have some defensive value – the animals would have made tasty food for larger fish and marine reptiles in the Mesozoic – the shells were quite exquisitely designed for controlled locomotion in the seas. That facilitated an active, predatory lifestyle that worked well for the ammonoids for a very long time. The shell design may have been more about catching food than avoiding being eaten. This knowledge of function – namely, that every ammonite fossil is the relic of a complex natural submarine, and one built by an invertebrate mollusk – can only make one marvel at the fossils more.

There is a large literature on the nature of biological functions, and I will not try to survey it here. But one of the leading accounts of biological function is the selective-historical theory, inspired by Wright (1973), given its canonical form by Millikan (1984, 1989), and further elaborated on by Godfrey-Smith (1994), Neander (1991), Preston (1998), and others. According to the selective-historical theory, the function of an item is something that it was naturally selected for doing in the past. Consider, for example, the function of perspiration in humans. Perspiration has lots of effects (like causing body odor), but its biological function is to help regulate body temperature. Perspiration contributed to the survival and reproductive success of our ancestors by helping them maintain a constant body temperature while staying active in hot environments. If this selective-historical theory is correct, then it means that knowledge of biological function is basically knowledge of history. Parsons and Carlson (2008), who claim that knowledge of function enhances aesthetic appreciation, are just drawing out one consequence of historical cognitivism.

One reason why it's important to see Parsons and Carlson's account of functional beauty as a special case of historical cognitivism has to do with nonfunctional traits. Gould and Lewontin (1979) famously argued that some traits are spandrels, which have no direct biological functions. In the basilica of San Marco in Venice, the dome rests on four arches, and where the arches come together to form corners, there are stunning mosaic panels depicting Matthew,

Mark, Luke, and John. But the spandrels – the parts of the structure where the mosaics are – were not put there for the purpose of holding the mosaics. They are, rather, inevitable by-products of placing a dome on four arches. Many traits are like that. The armpit, for example, is not really an adaptation for perspiration; it's just a by-product of having one's forelimbs attached to the torso in a certain way. Interestingly, though, the claim that a biological trait is a spandrel is *also* a claim about its evolutionary history. Knowing whether a trait is an adaptation *vs.* a spandrel can (again, on the assumption of historical cognitivism) affect our appreciation of it. I want to resist the suggestion that non-functional traits are somehow less beautiful than functional ones. The point, rather, is that the hypothesis that something is a spandrel – much like an adaptationist hypothesis – helps to guide and direct our aesthetic engagement with that thing.

Although some aspects of the functional morphology of ammonoids are fairly straightforward, others pose some puzzles. For example, during the Mesozoic (251–66 million years ago), there was a noticeable trend in the ammonoids toward greater shell complexity and ornamentation (Monnet, Klug, and de Baets 2015). Ammonoid shells from earlier in the Mesozoic tended to be smooth and rather boring. Later on, different types of ammonoids evolved a wide variety of ribs, nodules, and spines on their shells. What explains this trend toward greater ornamentation over time? One suggestion is that the trend was driven by natural selection (Ward 1981). The Mesozoic also saw the proliferation of big fish and mosasaurs (marine reptiles) that probably preyed on ammonoids. If a mosasaur wanted to eat an ammonoid, it would probably have to crush the shell in its jaws. Did the increased ornamentation of the ammonoid shells serve as a defensive response against shell-crushing predators? Maybe a shell with pronounced ribs, for example, has greater structural integrity. Or maybe the spines helped make the animal look bigger and less palatable to predators than it really was. Still another possibility is suggested by Dan McShea and Robert Brandon's (2010) notion of the *ZFEL*, or zero-force evolutionary law (pronounced like "zeffel."). McShea and Brandon think that in evolutionary systems, structural complexity just naturally tends to increase over time, and not necessarily as a result of natural selection. If this is right, then the trend toward increasing ornamentation in ammonoids might have nothing much to do with defense against predation. The trend might be an instance of the kind of complexity increase that "just happens" over the course of evolutionary history, unless some other factors keep it from happening (Turner 2018, pp. 327–9). How you perceive an ammonoid fossil – what you see the ribs and spines *as* – depends on which of these stories you think is most probably true. Different beliefs about function (or lack thereof) lead you to experience the fossil in different ways.

Theoretical morphologists are interested in both actual and possible biological forms. In a classic paper, paleontologist David Raup (1966) constructed what is known as a 3D morphospace for coiled shells, including things like snail shells and ammonoid shells. Think of a morphospace as a space of biological possibility, or in this case as a space of possible shell shapes. Each axis of the grid represents one measurable feature of the shell geometry. This limits things considerably, since it means that the morphospace can only represent three distinct traits. And there are lots of distinct traits or measures that we might be interested in. In order to represent the full sweep of morphological variation in shell design, we'd need an absurdly complex multidimensional morphospace, with one dimension for each trait we might be interested in – and that is obviously impractical (MacLaurin and Sterelny 2008). So Raup's original morphospace left lots of things out. He focused only on the geometry of shell coiling, including features such as expansion rate. The whorl expansion rate of a shell is the ratio of the diameters of two successive whorls. This approach left out lots of other features, including the ribs, nodules, and spines. Decisions about how to construct a morphospace in the first place may well involve aesthetic judgment, since they involve foregrounding and focusing attention on particular characteristics.[24] Constructing a morphospace brings out some interesting biological questions. We can think of evolution as following a path through the morphospace. But why did it take that particular path, rather than some other? Some parts of the morphospace are puzzlingly empty. Why did some possible shell geometries never get tried out? And why does evolution sometimes go "off the map," so to speak?

Late in their evolutionary run, a group of ammonoids, sometimes known as the "heteromorphs" (formally classified as the *Ancyloceratina*, a mash-up of Greek and Latin that means "crooked horns") seem to have busted out of the ordinary ammonoid morphospace. For millions and millions of years, most ammonoids (though not all) were variations on the classical coiled-shell theme. But during the Cretaceous period, new forms appeared with uncoiled shells, some of which were stretched out in weird, irregular forms, and some of which even looked like they were tied up in knots. These heteromorphs pose an aesthetic as well as an epistemic challenge. Aesthetically, they are a little bit like modernist orchestral music that challenges conventional ideas about

---

[24] In this connection, McLaurin (2003) and Currie (2012) draw a useful distinction between a *theoretical morphospace*, constructed a priori, and *an empirical morphospace,* constructed on the basis of data from real populations. In both cases, aesthetic considerations may play a role in determining which traits to focus on.

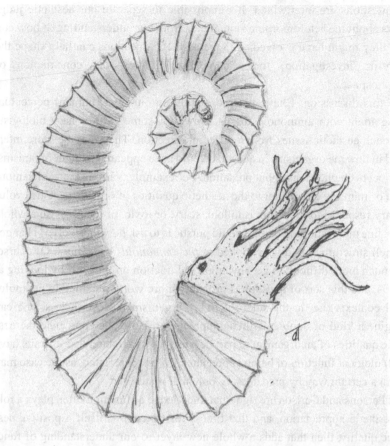

Figure 7 Heteromorphic ammonoid, suborder *Ancyloceratina*. Artwork by the author.

harmony and structure. The classic coiled shell form is obviously so beautiful in a formalist sense; when you look at heteromorph fossils, it may strike you that the ammonoids were going into (aesthetic) decline, or that something had gone badly wrong. The epistemic challenge is to figure out why some lineages departed from the classic coiled shell structure that had worked so well for so long. Did the heteromorphic structures provide some advantage? If so, what? One recent speculative suggestion is that the heteromorphs lived stationary lives attached to kelp (Arkhipkin 2014), though that would render their amazing chambered buoyancy system otiose (see Landman et al. 2014 for a critique of the kelp idea). Or could this be another instance of McShea and Brandon's *ZFEL* – a mere result of the blind tendency of structure variety to increase over time? I don't know the answers to these questions; the main suggestion is that the questions are intertwined. It's impossible to separate our aesthetic judgments about the heteromorphic ammonoids from our understanding of how and why they might have evolved. And the aesthetic questions can help shape the scientific investigation, for example, by guiding the construction of morphospaces.

In this discussion, I have focused mainly on our own (human) perceptual engagement with ammonoid fossils. However, some evolutionary biologists approach aesthetic issues from a different direction. They may be more interested in how the organisms in question would have appeared to other organisms. This bears on questions about predation. For example, various kinds of camouflage or mimicry can matter to the aesthetic qualities of organisms. One evolutionary response to predation is to look scary, or toxic, or inedible. So a whole other line of questioning that we could pursue is to ask how the evolved changes in shell structure would have looked *to the ammonoids' predators*. Of course, it's much more difficult to get any empirical traction on this, just by looking at fossils, and this sort of question is probably more worth pursuing in neontological contexts (i.e. in the study of living organisms). Nevertheless, one can imagine a kind of double aesthetic appreciation here. We appreciate the aesthetic qualities of an organism in part because we understand that its traits have the biological function of being appreciated (or unappreciated, as the case may be) in a certain way by predators, or potential mates.

If Parsons and Carlson are right that knowledge of function often plays a role in aesthetic appreciation, and that fitness-for-function is itself a positive aesthetic quality, then that adds a whole new layer to our understanding of functional morphology. Efforts to infer the functions of fossilized structures, such as the ornamentation of ammonoid shells, or the bizarre hooked forms of the heteromorphs, acquire a new aesthetic dimension. Because knowledge of function plays a role in aesthetic appreciation, one could say that one goal of the scientific practice is to better appreciate the fossils, *qua* aesthetic objects. The epistemic blurs into the aesthetic here; inferring the functions of the fossilized structures is not only a means to appreciating them aesthetically, but the research itself is an exploration of the aesthetic qualities of the fossils.

6 Explaining Historical Scientific Success

Having introduced some of the basic ideas of paleoaesthetics, I now want to explore how all of this bears on some traditional issues in the philosophy of science. A good place to begin is with Currie (2018), who offers the most sophisticated available account of how historical scientists overcome various epistemic and methodological challenges. Currie zeroes in on what he calls "epistemically unlucky circumstances," or situations in which scientists attempting to reconstruct the past have few historical traces to go on, where the evidence is gappy, and where the historical signal it carries is faint. These are the sorts of cases in which you might expect scientists to confront severe underdetermination problems (Turner 2005b). In epistemically unlucky circumstances, the evidential traces might seem insufficient for discriminating between rival historical hypotheses. Currie, however, argues that in some cases, scientists really are able to make progress, and to draw solid inferences from fairly scanty evidence. He then sets out to explain how scientists achieve epistemic success in such unlucky circumstances. Currie takes a pluralist view of epistemic goods, so that "epistemic success" could mean achieving several different things: knowledge, true theories or hypotheses, accurate representations, better models, better explanations, and so on (compare also Potochnik 2015). Currie wants above all to make a case for optimism about historical scientific research. If scientists can make progress even in epistemically unlucky circumstances, then – once we understand *how* that progress gets made – we have some reason to be optimistic about historical science.

Currie takes a dim view of the traditional scientific realism debate, which raged during the 1980s and 1990s and continues in some quarters today. (Psillos 2018 offers a helpful overview, and see Miyake 2018 for an interesting defense of realism in the geosciences). We can understand Currie's view a bit better by juxtaposing it with more traditional scientific realism. One difference is that traditional realists are a bit less pluralistic about the aims of science. Whereas Currie, following Alison Wylie (1999), sees historical scientists pursuing a variety of epistemic aims, most realists would take the narrower view that truth (or maybe approximate truth) is *the* aim of science. Realists also tend to be more theory-centric, focusing on hypotheses and theories as distinctive products of scientific work. There are, however, a couple of similarities between Currie's view and classical scientific realism. Like the realists, Currie is interested in explaining why historical science is so successful. "That science progresses," he writes, "is almost a datum – something to be explained" (Currie 2018, p. 310). And like realists, Currie is generally optimistic about historical scientists' ability to generate knowledge, even of things in the deep past that we cannot observe. Indeed, some realists closely identify realism with epistemic optimism. For instance, Juha Saatsi writes that "a scientific realist defends a degree of rationally justifiable optimism regarding scientific knowledge, progress, or representational adequacy with respect to directly unobservable features of reality ..." (Saatsi 2018, p. 1). And Currie himself, in spite of voicing reservations about the types of arguments that traditional realists and antirealists have deployed, somewhat grudgingly concedes that his view places him at least in the vicinity of the realist camp (p. 321).

Although Currie shares with traditional realists an interest in explaining the success of science, there are important differences between his view and the more traditional framings. Traditional realists, following Hilary Putnam's famous "no miracles" argument, have focused on the success of theories, and defined success rather narrowly as empirical or observational success (Wray 2018).[25] A successful theory is one that gets all the observable phenomena right, a theory that generates accurate predictions. The realist's next move is to argue that this empirical success would be a miracle if the theory in question were not true or at least approximately true. "Why is theory T so successful at generating accurate predictions?" "Because it's true." This idea that truth explains success is the central component of most traditional versions of scientific realism.[26] Currie, however, has little interest in the "no miracles" argument. Like many philosophers concerned with scientific practice, he shifts the focus from the products of science (e.g. theories) to investigative processes. And he takes a much looser view of success. Success, for him, is just the achievement of any of a number of different epistemic goods, including truth. And unlike traditional realists, Currie makes no appeal to truth as an explainer of success. Instead, he identifies several aspects of the practice of historical science that conduce to success in epistemically unlucky circumstances.

Currie points to three particular features of investigative practice in the historical sciences that he thinks contribute to success: (1) methodological omnivory; (2) epistemic scaffolding; and (3) empirically grounded speculation (Currie 2018, p. 309). In defending methodological omnivory, Currie is responding critically to other philosophers, especially Carol Cleland (2002, 2011), who have suggested that historical science has its own distinctive method. Cleland, for example, argues that prototypical historical science involves the formulation of rival hypotheses that offer potential explanations of some collection of traces. Then scientists test those hypotheses by searching for what she calls a "smoking gun" – a trace or set of traces that helps scientists to discriminate between the rival historical hypotheses. However, not all work in paleontology fits Cleland's

---

[25] Some classic discussions of the "no miracles" argument and its descendants include Boyd (1984, 1990), Leplin (1997), Psillos (1999), and Putnam (1978). See Turner (2007, chs 3 and 5) for further discussion of the argument in connection with historical science. The literature on the "no miracles" argument and various critical responses to it (e.g. Laudan 1981) is vast and technically sophisticated, and I will not try to summarize it here.

[26] Rossetter (2018) offers a really sophisticated look at a case study from the history of geology with a focus on this issue of novel predictive success.

description (Turner 2009). Drawing upon Alison Wylie's (1999) suggestion that certain sorts of disunity actually contribute to successful inference in archaeology, Currie goes further in arguing that it's the lack of commitment to any particular methodology that helps explains historical scientists' successes. For example, when faced with a dearth of evidential traces, Currie thinks historical scientists can gain additional traction by expanding their toolkits to include various kinds of comparative methods, modeling, and experimentation.

Currie's notion of epistemic scaffolding is also quite useful.[27] As the metaphor suggests, scaffolding is any scientific work that has indirect pay-off for investigating or understanding something else. Building fancy scaffolding is a feat in and of itself, but the really exciting thing is what the scaffolding enables you to do. There are many examples of scaffolding in science – and Currie discusses quite a number – but one that we've encountered already in Section 5 is David Raup's morphospace for coiled shells. That morphospace is really just a representational tool; it idealizes away from many different features of actual shells, providing only an abstract characterization based on three traits. But it is a representational tool that helps to frame new questions and open up new investigative spaces that would otherwise be difficult to access.

Currie defines speculation as "the practice of making claims that go beyond available evidence by some relevant margin" (Currie 2018, p. 287). It might be tempting to take a more conservative view, and insist that scientists should keep their speculation to a minimum. But Currie argues that too much epistemological fastidiousness could actually leave scientists worse off. That's because speculation sometimes has difficult-to-predict, indirect, investigative payoff. He writes:

> Some hypotheses are justified on the grounds of their support—that is, whether we think they are true. Other hypotheses speculative ones, are justified on their fruits . . . (2018, p. 288).

Of course, Currie is not quite saying that anything goes. Speculation about extra-terrestrials causing mass extinction events in the deep past is not likely to be too fruitful. But sometimes, when scientists find themselves in epistemically unlucky circumstances, a speculative claim might open some surprising investigative doors.

To summarize, Currie thinks that methodological omnivory, scaffolding, and occasionally fruitful speculation all help to explain how scientists make progress in epistemically unlucky circumstances. These are features of the

---

[27] For other helpful discussions of scaffolding, see Chapman and Wylie (2016) and Walsh (2019).

investigative practice that explain success. You may already have guessed that I think Currie's account of historical scientific success – like virtually all the work done in the philosophy of historical science up to now, including my own – suffers from a serious limitation. It focuses on the epistemological dimension of historical reconstruction, as if that were the whole story. It leaves out the aesthetic dimension. This is a problem for two different reasons, which I elaborate below. First, it affects the characterization of scientific success. Second, it means that Currie's explanation of the successes of historical science, though largely correct, is incomplete. What I propose here is an aesthetic expansion of Currie's account of the practice of historical science.

Realists claim that the goal of science is truth or approximate truth. Some antirealists have famously challenged this. For example, Van Fraassen (1980) argued that the goal is merely empirical adequacy, or getting theories with true (and only true) empirical consequences. If I am right, this whole axiological debate between realists and antirealists is hopelessly narrow, because historical sciences like paleontology have other distinctively aesthetic goals as well. These aesthetic goals include better aesthetic appreciation of fossils and landscapes, the cultivation of sense of place, and even the production of various kinds of artwork: 2D paleo-artistic representations, prepared fossils, museum exhibits, and so on. Currie is much more broad-minded than traditional contributors to the realism debate, allowing for historical science to have a variety of (often localized, fine-grained) epistemic aims and goods. He rightly resists both the traditional realists' and antirealists' grandiose claims about "the" aim of science. Nevertheless, when Currie gets around to specifying what epistemic goods he has in mind, his list looks pretty traditional. The "outputs" of scientific investigation, he says, include "truth, understanding, predictions, prognostications, and so forth" (p. 281). But if we follow Currie in adopting a more pluralistic, ecumenical view of the goods of scientific research, it's hard to see how there could be any principled reason for narrowing our vision to include only the epistemic goods. A genuine axiological pluralism should be open to the possibility that science has some important non-epistemic (especially aesthetic) goals as well as epistemic ones.[28] Or to approach this from another direction, Currie is already defending a quite expansive pluralism about the methods, tools, or means of historical science. He appropriately criticizes other

---

[28] Potochnik (2015) gestures in this direction but does not explicitly include aesthetic goals in her conception of the diverse aims of science.

philosophers who take historical researchers to be "methodological obligates" committed to using just one basic approach to reconstructing the past. But he still treats historical scientists as "axiological obligates." Like organisms that can only eat one type of food, he sees scientists as pursuing just one type of aim – namely, the epistemic ones. If you think of an omnivore as someone who eats meat as well as plant-based foods, it might make sense to say that historical scientists are axiological omnivores, too; they pursue aesthetic as well as epistemic goals. This way of putting things does presuppose the distinctness of aesthetic and epistemic aims. If those blur together (as suggested by the blurring view), then the pursuit of epistemic aims is also and at the same time the pursuit of aesthetic aims.

Once you allow that historical science may have aesthetic as well as epistemic goals, that further complicates characterizations of success. Success is no longer merely a matter of having theories that generate accurate predictions, or even a matter of gaining knowledge of the deep past. Aesthetic success, too, is something that needs to be explained, just as much as epistemic success. Of course, if historical cognitivism is correct, then epistemic success could well be part of the explanation of aesthetic success.

In addition to expanding our notion of scientific success, paleoaesthetics points to the need for taking a more expansive view of how to explain local scientific successes. Here Currie's notion of methodological omnivory is especially helpful, because it opens the door to thinking about a wide variety of investigative practices and tools that can yield epistemic results. The main point I wish to make is just that those investigative practices need not be construed as narrowly epistemic. Currie's own discussion of paleoart (Currie 2017b) moves us in this direction. He argues that good paleoart can actually be a part of the scientific process, because it can sometimes depict bold conjectures about prehistoric life – note the connection to Currie's notion of pragmatically justified speculation. Good artwork can be a way of floating new scientific trial balloons that go a bit beyond available evidence, while remaining tied down by it, and in doing so the artwork can help to clarify new questions and inspire us to identify new lines of evidence. One excellent example of this is the artwork in *All Yesterdays*, by Conway, Kosemen, and Naish (2012). They offer speculative representations of many familiar dinosaurs, representations that are constrained but not dictated by the fossil evidence.

Currie's defense of paleoart focuses on a particular genre, more or less speculative 2D representations of dinosaurs and other prehistoric creatures that portray them as living animals, usually in their imagined environments. I think Currie is right that this artistic genre has always played an important role in paleontological research. But we can go much further. Paleoart is a type of artistic work that many of us nonspecialists see all the time – in popular dinosaur books, museums, and even postage stamps. So it is a natural place to begin. But if we really want to understand the success of the historical sciences, we also need to consider other aesthetic practices, such as geological field sketchwork.

Although this may be changing with increasing reliance on digital photography, the production of field sketches has long played a central role in the practice of the earth sciences. Figure 7 offers just one example of a field sketch, from G.K. Gilbert, a geologist and a fabulous artist who accompanied John Wesley Powell in his famous survey of the Rocky Mountain region from 1874 to1879. This sketch is from Gilbert's 1877 work, *Geology of the Henry Mountains*. It could be an ordinary bit of landscape art, but Gilbert has carefully positioned himself so as to be able to depict an unconformity, at a site near Salina, Utah. The horizontal strata above are Paleogene rocks, while the inclined strata below are from the

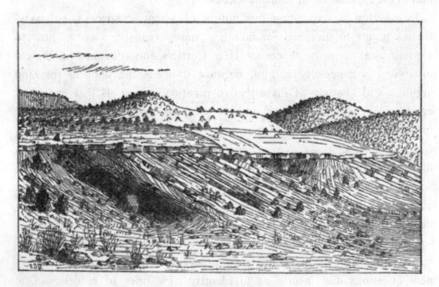

Figure 8 Geological field sketch from C.K. Gilbert depicting an unconformity near Salina, Utah. The lower inclined strata are Cretaceous rocks. The upper, horizontal strata represent the Paleogene. Public domain image, courtesy of the US Geological Survey. From G.K. Gilbert (1877), *Report on the Geology of the Henry Mountains*. US Geological Survey Unnumbered Series. Washington, DC: US Government Printing Office, p. 15.

Cretaceous. In short, he is showing us the K-Pg boundary. And notice how the composition is crafted so that the crucial geological boundary bisects the frame horizontally.[29]

A field sketch like Gilbert's is rather different from the genre of landscape drawing or painting, although they have some things in common. One crucial difference is that a field sketch involves more abstraction. A map or diagram always includes some details while leaving out others.[30] For example, a public transit map might include subway lines and bus stops, while leaving out streets and elevation changes. Similarly, a geological field sketch includes some information about the landscape while leaving out lots of detail. It might include representations of rock strata and elevation, while leaving out details about foliage cover and local wildlife. Another difference between field sketches and landscape art is that sketches involve lots of textual labels and annotations; in this they are again much more like maps. One main purpose of a field sketch is to depict local geological structure. Historically, this was a way of recording information about a site; sketches could then serve as data for geological theorizing at larger scales. There are also some important similarities between geological field sketches and landscape art. For example, the geological sketch artist must also make a decision about how and where to situate herself in the environment; every field sketch, like every landscape painting, is carried out from a particular perspective.

The skills involved in producing a good field sketch are very similar to those involved in drawing a landscape. Geological sketchwork is pretty obviously an artistic practice. And it has functions that go beyond the mere recording of information about a site. Drawing, in general, is a way of developing and honing one's perceptual skills. The practice of drawing people's faces, say, will, if nothing else, require sustained and careful perceptual attention to details that many of us do not ordinarily focus on too much. For example, a good artist may develop the skill of quickly noting the shape of someone's face – is it a bit more oval, or a bit squared off? In this way, the material practice of drawing, with pen and paper (or with whatever tools), feeds back into perception and can make a difference to how and what one notices. Thus, in addition to the direct benefit of recording information about an outcrop, geological sketchwork has profound indirect benefits of helping researchers develop and maintain the skills

---

[29] I thank Rob Inkpen for calling my attention to Gilbert's exquisite geological drawings.

[30] Geological mapping is another semi-artistic practice that's crucial to the success of the paleosciences, and one that deserves more attention than I can give it here. See Oldroyd's (2013) account of the history of geological mapping, with attention to some of the ways in which mapping became more diagrammatic. And of course, digital technology is transforming these practices in ways that deserve much more attention than I can give them here.

necessary for "seeing" the geological structure of the landscape. The practice of sketching geological structure by hand arguably deepens one's aesthetic engagement with the land. But – and this is now the linchpin of my argument – the larger empirical successes of geology and paleontology would be utterly inconceivable without building on this long tradition of artistic practice. It's hard to imagine how the field of stratigraphy could have developed at all without field sketches. If nothing else, geologists needed some way of sharing information about sites they were theorizing about.

We philosophers tend to think of knowledge as propositional, and so we think of scientific knowledge as involving propositional structures (networks of theories and hypotheses). But this view is arguably too narrow (Langer 1942). A visual representation, such as Gilbert's field sketch, is many things at once. It's an epistemic product – a bit of scientific knowledge – as well as a representational tool that other geologists can put to work in various ways. It's also an aesthetic product, an artwork. Because one and the same thing is serving as an epistemic tool, epistemic product, and an artistic product, this is not merely a case in which artistic skills are being deployed in the service of epistemic aims. It's a case where artistic and epistemic practices tightly interweave, and may even blur together.

Note that field sketches have very different scientific functions from paleoart. Paleoart is largely – though as Currie argues, not entirely – for consumption by the nonspecialist general public. Field sketches are more for internal scientific use. All paleoart involves some degree of speculation, an imaginative rendering of prehistoric scenes that no one could have observed. But field sketches are not essentially speculative in this way. Like maps and diagrams of all sorts, they are depictive tools. The practice of drawing field sketches is an aesthetic practice if anything is, but it's one that has contributed greatly to the epistemic successes of geology and paleontology.

In this section, I've argued that paleoaesthetics matters a great deal for one traditional project in the philosophy of science. Traditional realists appealed to truth to explain the empirical success of theories. And today, philosophers like Adrian Currie, who has tried to distance himself from the realism debate, nevertheless remain in the business of trying to explain the successes of science. But if the line of argument I've been developing is on the right track, these efforts to explain the success of science are too narrow in two different ways. First, their construals of the aim(s) of science are too narrow. And second, they neglect aesthetic aspects of the practice of science, such as the production of field sketches, that are highly relevant to scientific success. The trouble with scientific realism is not that it's false, but that it offers a one-sided construal of science, privileging theory while neglecting practice, and completely overlooking the aesthetic dimensions of natural science. Canonical versions of anti-realism, such as traditional instrumentalism, or Van Fraassen's (1980) constructive empiricism, just inherit these deficiencies.

7 Fossils as Epistemic Tools and Aesthetic Things

One valuable service that we philosophers of science can provide is the critical examination of scientific metaphors. No metaphor is good or bad, full stop. Metaphors are more or less generative; they open up new lines of questioning and new ways of seeing, but once a metaphor becomes entrenched, it can also prevent us from seeing certain things in certain ways. Once you start looking for metaphorical concepts in science, you see them everywhere. Metaphors structure our thinking in ways that we do not always fully appreciate (Larson 2014; Turner 2005a). And it goes without saying that metaphor is the stuff of poetry. Once you appreciate the ubiquity of metaphorical concepts in science, it's hard not to see scientific concept formation and theory-construction as semi-poetical practices. Indeed, Mary Hesse (1966) argued that scientific theories explain via metaphorical redescription of target phenomena. The importance of metaphor to science is also a long-running theme of some of Michael Ruse's work (see, e.g. Ruse 2005, 2013). Stephen Jay Gould (1988) and David Oldroyd (2006) have explored the importance of the cycle metaphor to the development of geological thought. Finkelman (2017) discusses the use of spatial metaphors for thinking about geological time.

If the centrality of metaphor is less than obvious, consider this short list of metaphorical concepts:

> Living fossil
> Plate tectonics
> Random genetic drift
> Mantle plume
> Cambrian explosion
> Spindle diagram
> Gene flow
> Chemical bonds
> Snowball earth
> Invasive species
> The big bang
> Molecular clock
> Phylogenetic tree
> Ecological niche
> Cell wall
> Genetic transcription and translation
> Biotic community
> Etc.

If you still have doubts, think a bit about the names that we give to extinct species: *Tyrannosaurus* ("tyrant lizard") or *Maiasaura* ("Good mother lizard"). Those are evocative – and gendered! – metaphors that say more about us than they do about the animals so named.[31] Or recall the ammonoids, whose name harkens back to the Egyptian god Ammon, who had horns like a ram.

One worthwhile intellectual exercise is to start with a familiar metaphor – say, "ecological niche" – and think of alternatives to it:

> Ecological pedestal
> Ecological office
> Ecological basket

Each of these has slightly different connotations, and those different connotations can make a theoretical difference. For example, it's natural to suppose that only one thing can occupy a "niche" or a "pedestal" at one time. But multiple things could be placed in a basket or an office. It is relatively easy to knock something off of a pedestal, but a bit tougher, perhaps, to dislodge something from its niche. Owing to these subtle connotative differences, a change in metaphor could make for a change in ecological theory. Every scientific metaphor involves a kind of collective conceptual decision, and as in poetry, there are always other options. Moreover, once such a decision gets made, the metaphor can shape our thinking in ways that are sometimes very difficult to notice.

One metaphorical concept that has profoundly shaped our thinking about paleontology is the *fossil record* – or perhaps a bit more broadly, the *geological record*. At the heart of this metaphor is the idea that the crust of the Earth is like a book or text, waiting to be interpreted.[32] To get a sense for just how strongly this metaphor shapes our thinking about paleontology and geology, consider the titles of just a few books: *Written in Stone: A Geological History of the Northeastern United States* (Raymo and Raymo 2001); *The Meaning of Fossils* (Rudwick 1972); *Annals of the Former World* (McPhee 2000); *Written in Stone: Evolution, The Fossil Record, and Our Place in Nature* (Switek 2010); *Rereading the Fossil Record* (Sepkoski 2012); and *The Rocks Don't Lie: A Geologist Investigates Noah's Flood* (Montgomery 2012).The common

---

[31] Interestingly, there are only a handful of dinosaurs with the feminine "-saura" suffix. *Leaellynasaura* is another one.

[32] But see also Smith (2019), who isn't convinced that this is even a metaphor. Smith takes a much more expansive view of what counts as "reading," so in his view, interpreting fossils could literally count as reading. This move is generally available: Suppose you start with a metaphor, such as "Her mind is a steel trap." You can broaden the definition of "steel trap" so that the claim becomes literally true. Just define "steel trap" as anything that rapidly grasps and holds on to things.

theme here is that the fossil record is a book, or an annal, or something written, something to be read, something with meaning that needs to be sussed out and interpreted, something that could testify truthfully about the past. The notion that fossils constitute a textual record is so familiar that we may not even think of it as a metaphor, but of course literally speaking, fossils are not a text, and they are (literally) nothing readable or re-readable.

It might seem that the notion of a fossil record is totally innocuous. I do not want to argue that the metaphor is a bad one or that we should give it up. My claim is the much more modest one that the metaphor structures how we think about fossils, and paleontological investigation, and that the textual metaphor has had the effect of making the aesthetic dimensions of paleontology harder to see. Like any other metaphor, the fossils-as-text idea has cognitive costs and benefits. Rudwick (2016) shows how the textual metaphor contributed in a positive way to the development of geology, as early geologists borrowed and extended some of the methods of text-based chronology. Virtually all work in the history and philosophy of the paleosciences so far has pretty much taken this metaphor as a given, without pausing to consider other possible ways of thinking about fossils. To give just one example, Robert Frodeman has argued that geology is an interpretive or hermeneutic discipline (Frodeman 2003, 2014; Raab and Frodeman 2002). He draws on the hermeneutic tradition in continental philosophy, which in turn drew inspiration from the development of sophisticated methods of textual analysis and criticism by philologists and Biblical scholars in the nineteenth century. Frodeman's treatment of geology as a hermerneutic science reflects his commitment to the textual metaphor.

If we really want to see why the textual metaphor matters, we would do well to look to Charles Darwin. In the *Origin of Species*, Darwin notoriously sought to insulate his theory from criticism by appealing to the incompleteness of the fossil record. He knew that his theory seemed to imply that there should be transitional forms in the fossil record. For example, if tetrapods (including amphibians, reptiles, birds, and mammals) evolved from fish or fish-like ancestors, then you might expect to find fossil specimens with a mix of fishy and amphibian traits. As of 1859, however, naturalists in the field had not turned up any unambiguous cases of transitional forms. The famous discovery of *Archaeopteryx* in the Solnhofen quarry in Bavaria would not happen for a few more years. Anticipating objections, Darwin attributed the lack of evidence for his theory to the incompleteness of the geological record. And in doing so, he leaned heavily on the textual metaphor:

> For my part, following out Lyell's metaphor, I look at the natural geological
> record, as a history of the world imperfectly kept, and written in a changing
> dialect; of this history we possess the last volume alone, relating only two or
> three countries; of this volume, only here and there a short chapter has been
> preserved; and of each page, only here and there a few lines (1859/1964, pp.
> 310–11).[33]

On the one hand, the textual metaphor lends itself to thinking about the fossil
record as a form of testimonial evidence about the deep past. Darwin is here also
using the textual metaphor to guide our thinking about the quality of that
evidence: It's gappy (because we are missing whole volumes) and difficult to
decipher (because written "in a changing dialect"). The textual metaphor thus
helps to capture the ideas that only some information about the past gets
recorded in the first place, and that once recorded, historical processes also
tend to degrade and destroy information, like bookworms eating through docu-
ments in an archive. The overall impact of the textual metaphor is to structure
our thinking about fossils in evidential terms. The dominant questions then
become: How much can the incomplete fossil record tell us about the past? How
much information can we extract from it? These questions have, in their turn,
led philosophers to argue about how much information gets preserved *vs.*
destroyed. This is a point of disagreement between Cleland (2002) and Turner
(2005b), as well as the subject of information-theoretical modeling by Sober
and Steel (2014). Adrian Currie (2018) tries to swap out the old metaphor with
his "ripple model" of historical evidence, but the basic questions he is addres-
sing with his ripple model are those that were implicit in the textual metaphor all
along.

This evidential understanding of fossils as a textual record is so compelling
that we may even be tempted to treat "fossil" itself as an evidential concept
(Turner 2011, ch. 10, explores this possibility). We could, for example, define
a fossil as any present trace that, in conjunction with certain background
theories, "tells us something" about past life. This is similar to how Currie
(2018, p. 70) defines "traces."

The textual metaphor has a history that goes back well before Darwin and
Lyell, to the early modern idea that God – often referred to in this connection as
"the Author of Nature" – had written two books for us to study. On the one hand,
there is scripture, the text of revealed religion. On the other hand, there is the
"book of nature" which serves as the focal point of natural theology. This two

---

[33] Darwin cites Lyell, who had used the same metaphor in a slightly different way, in his *Principles
of Geology*. In reference to pre-Christian thinkers, Lyell says that "the ancient history of the globe
was to them a sealed book, and, although written in characters of the most striking and imposing
kind, they were unconscious even of its existence" (Lyell 1853, p. 16).

books idea shows up in the work of many thinkers in the 1600s to 1700s, from Galileo to Joseph Butler. Indeed, the whole point of natural theology was to treat the natural world around us as evidence of God's existence and nature. By reading the book of nature, we can learn something about its ultimate author, and in this way, the textual metaphor provided the framework for subsequent developments of the argument from design.[34] Our common expression, "That is/isn't set in stone" probably refers to the Ten Commandments – the paradigm case of revealed religion. On the two books view, God has a penchant for writing in stone, whether it's on stone tablets shared with Moses (Exodus 24: 12), or in the very crust of the Earth.

During the nineteenth century, paleontology and geology gradually became more secularized, though with lots of fits, starts, and reversals. The idea that nature is a text persisted, even as science got out from under natural theology. The concept of the fossil record retains much of this heritage. It invites – even pushes – us to think of fossils chiefly as evidential items. There is nothing necessarily wrong with this framing. Fossils *are* evidential items, and it would be crazy to suggest otherwise. But like any metaphor, the "fossil record" concept is optional. Other metaphors, none of which are perfect, can open up new ways of conceptualizing things, and can bring the aesthetic dimensions of paleontology into clearer view.

One alternative might be to think of fossils as investigative tools.[35] Perhaps the fossil record is more like the fossil toolkit. This might seem a bit counterintuitive at first. When we think of scientific tools, we tend to think of equipment or technology: rock hammers, plaster, glue, drills, brushes, CT scanners, 3D printers, databases, and so on. But one important lesson from Caitlin Wylie's work is that the processes of collection and preparation involve a lot of human artifice. The prepared fossil specimens that result from those processes are aptly characterized as tools that pre-parators shape and modify for investigative purposes. The preparation process involves removing material to reveal something that you can put to use. The fossil record metaphor and fossil toolkit metaphor provoke different questions. If we think of fossils as tools, the question of complete-ness – the question that Darwin obsessed about – does not even come up, at least not in its familiar form. Instead, the salient questions are: What can we do with these tools? How can we put them to work in combination with

---

[34] For example, in Hume's *Dialogues Concerning Natural Religion*, Cleathes explicitly frames the design argument as involving an inference from the existence of books to the existence of an author (Hume 1988, p. 24).

[35] This is partly inspired by Ken Waters's (2008) argument that the success of twentieth-century genetics was largely a matter of finding new ways of treating genes as tools for investigation.

other tools? Are these available tools sufficient for doing what we want, or do we need to think about developing new ones? We could even define "fossil" in a pragmatic spirit, as any tool for investigating prehistoric life that is acquired via collection and preparation. Instead of reading (or rereading) the fossil record, we can think of paleontologists as developing and refining the fossil toolkit, and putting it to use in coordination with other investigative tools.

This notion of an investigative tool is still largely an epistemic one, since the idea is that we use fossils as tools for certain epistemic aims, so it might not be totally clear how this proposed metaphorical change-up does much to foreground the aesthetic dimensions of paleontology. So I want to conclude this section with a suggestion inspired by the work of Hans-Jörg Rheinberger (1997). In an effort to analyze the material practice of experimental science, Rheinberger draws a helpful distinction between "technical objects" and "epistemic things." The epistemic things are the targets or objects of investigation, which may be only vaguely delimited or understood at the outset – roughly, the items, processes, or systems that scientists want to learn something about. The technical objects of science are the various devices and tools used in the investigation – the lab equipment, measurement devices, aids to observation, and so on. As we learn more about epistemic things, they can then be deployed as technical objects for new investigations. One appealing thing about Rheinberger's analysis is his emphasis on scientists' concrete interactions with things in the world. Both the tools and targets of science are material things. Now although Rheinberger's work, like that of so many others, still exhibits something of a bias toward the epistemic, this bias isn't really essential. We could just as easily talk about *aesthetic things* – items, processes, or systems that are targets of aesthetic engagement or appreciation. And of course, if science is an aesthetic practice, it must have aesthetic as well as epistemic things. At this point, one of my central claims comes (back) into focus. Fossils are both "technical objects" of paleontological research – part of the paleontologists' toolkit – and aesthetic things of paleontological science. Understanding this dual role that fossils play is crucial for understanding the practice of paleontology. But really absorbing this might mean that we have to extricate ourselves from the grip of the idea that fossils are primarily a "record" of the past.

8 The Dinosaur Phylogeny Debate

At this point, you might be convinced that there is something going on in paleontology that we could loosely call "aesthetic." Maybe you're convinced that scientists do have an interest in engaging aesthetically with fossils and with landscapes. Maybe you even find historical cognitivism to be rather plausible. But you might still be skeptical about the centrality of all these aesthetic considerations. Ultimately, you might think, paleontology (or any science, for that matter) is epistemic at its core. Science is about learning stuff, or about finding things out. The aesthetic dimension might be interesting and real, but peripheral and inessential. In this section, I will try to drive the argument home with a case study. Surprisingly, it turns out that one of the biggest and liveliest ongoing debates about dinosaurs boils down (at least at the moment, as of this writing, though this could change) to different scientists' aesthetic engagement with a single fossil, like oenophiles giving somewhat different assessments of a fine wine.

I want to be careful not to make too much of this example. The controversies that crop up in dinosaur science, where sample sizes are often quite small, may differ quite a bit from disagreements in other areas of paleontology with different empirical questions, methods, and standards. So I do not want to claim that this example is representative. This is also, in some ways, a more traditional problem of how to decide between two rival hypotheses. But I do offer it as a case where it turns out that aesthetic engagement is right at the core of scientific activity. I think this shows that sometimes the aesthetic considerations are not peripheral at all. And unlike traditional cases of theory choice, this is not one where the scientists are assessing the aesthetic qualities of the rival theories; in this case, rather, the aesthetic thing is a single, confusingly fragmentary skeleton that might or might not be a dinosaur.

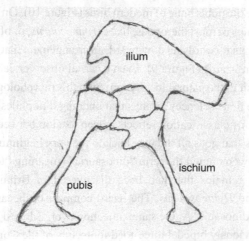

Figure 9 Saurischian (lizard-like) hip construction.

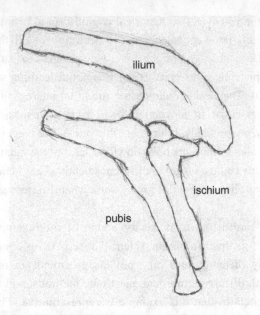

Figure 10 Ornithischian (bird-like) hip construction.

Traditionally, dinosaurs have always been classified as belonging to either of two groups: the lizard-hipped saurischians or bird-hipped ornithischians. This distinction goes back to the late nineteenth century, and thus long predates contemporary cladistic approaches to classification. Originally, the distinction was just based on morphology. If you look at the hips of your favorite ornithischian dinosaur – say, *Stegosaurus*, or *Triceratops*, or any of the "duck-billed" hadrosaurs – the pubis bone points aftward in the direction of the tail, and that looks a lot like the pubis bone of modern birds (Figure 10). On the other hand, if you look at the hips of most theropods, like *Tyrannosaurus*, or of the sauropods, the pubis bone points forward and downward, an arrangement that looks a lot like the hips of modern lizards (Figure 9). Even a casual observer can easily tell the difference on a visit to the natural history museum; this morphological difference is often one of the first "sciencey" things that budding dinophiles learn as kids.

This dichotomous classification reflects a deep tension between two ways of viewing dinosaurs that goes all the way back to the very beginning. In the early 1840s, Richard Owen coined the term "dinosauria" (meaning "terrible lizard") to cover skeletal remains that had been discovered in Britain: *Iguanadon, Megalosaurus*, and *Hylaeosaurus*. The lizard comparison became entrenched in the very term "dinosaur." At the same time, however, Edward Hitchcock was attributing the obviously bipedal three-toed footprints of the Connecticut River Valley in North America to gigantic prehistoric birds. Not only did he not

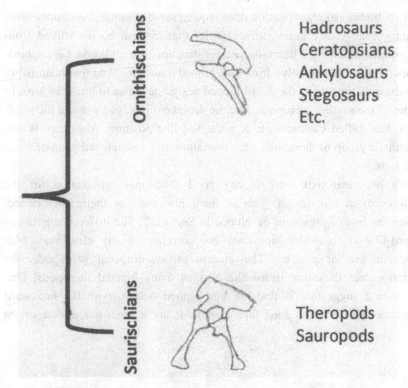

Figure 11 The traditional phylogeny of the dinosaurs.

classify them as dinosaurs, but he supposed that his avian "Brontozoa" must have lorded over the lumbering British lizards, which everyone at the time thought were quadrupedal.[36] This clash of perspectives has cropped up over and over again – for example, whether dinosaurs were warm-blooded or cold-blooded was very largely a question of whether they were more bird-like or lizard-like. And the ambivalence about which living creatures to use as models for dinosaurs is reflected to some extent in the ornithischian/saurischian distinction.

Until very recently, the old ornithischian/saurischian division had persisted through the adoption of cladistic methods. The rough idea was that during the middle Triassic period (251–199 million years ago), dinosaurs split into these two groups, with the earliest theropods and sauropods taking one evolutionary path, and everyone else going a different way. In this view, these two groups represent distinct clades, readily diagnosable by hip morphology. (A clade is a section of the Darwinian tree of life defined by an ancestral group, plus all and only those species that descended from it.)

---

[36] For more on this, see Turner (2017a).

This traditional classification does require some assumptions about evolutionary change. We know, for example, that modern birds evolved from theropods, which had a lizard-hipped construction. It sounds odd, but no, bird-hipped birds did not evolve from bird-hipped dinosaurs. And that is probably not the only time when the lizard-hipped design gave rise to bird-like hips. In some of the smaller theropods, like the dromaeosaurs, you can see the pubis bone has shifted backward, in a more bird-like position. And there is one enigmatic group of theropods – the therizinosaurs – which had genuine bird-like hips.

It's not clear that there is any good functional explanation for the differences in hip design. This in itself has some aesthetic importance, given the line of argument developed in Section 5. The difference between lizard-like *vs*. bird-like hips does not correlate in any clear way with dinosaur gait or posture. The ginormous, quadrupedal sauropods, for instance, had the same lizard-like hips as many bipedal theropods. One interesting suggestion is that the bird-hipped design, with the backward pointing pubis bone, is good for accommodating a larger gut, and so might

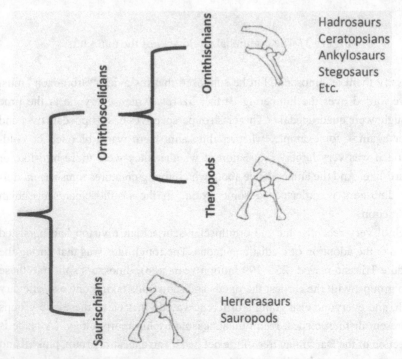

Figure 12 The new dinosaur phylogeny proposed by Baron, Norman, and Barrett (2017a).

have been selected for in herbivorous lineages. This might help make sense of the fact that the therizinosaurs, those anomalous herbivorous theropods, had bird-like hips. But it doesn't explain why some big quadrupedal herbivores, like the ceratopsians, had bird-like hips, while others, such as the sauropods, retained lizard-like hips. In a word, the traditional ornithischian/saurischian classification, deeply entrenched as it is, raises lots of unanswered questions about dinosaur evolution.

Recently, Baron, Norman, and Barrett (2017a) published a paper that upended this traditional taxonomy. They undertook a new phylogenetic analysis, using over 400 characters from dozens of dinosaur species. Importantly, they looked at many traits having nothing to do with the hip joint. They argued that there is no terribly good reason to fixate on the hip joint as being an especially important diagnostic character. And their more expansive phylogenetic analysis yielded a result that conflicts with the traditional view. Baron, Norman, and Barrett found that the theropods (*T. rex, Velociraptor*, and kin) are more closely related to the ornischischian dinosaurs than they are to the sauropods. Reviving an old nineteenth-century taxonomic category, they proposed a new clade called "Ornithoscelida" ("bird-limbed") that would include all the traditional ornithischians plus the theropods. The sauropods were not quite left out on their own, as they got placed in a clade with the herrerasaurs, a less well-known group of carnivorous dinosaurs. This research made something of a splash because it overturned one of the most basic things that we all thought we knew about dinosaurs. Baron, Norman and Barrett's relatively large sample size was a clear virtue of their work. However, as we'll see in a bit, it also turns out that disagreement about what to make of a single specimen in their sample can have a big impact on the phylogenetic results.

The new picture defended by Baron, Norman, and Barrett (2017a) has different implications for dinosaur evolution. For example, it seems to imply that the ancestral state, back during the early to mid-Triassic period, was the lizard-hipped design. The sauropods and herrerasaurs retained that design. Within the ornithoscelidan group, some of the theropods retained that ancestral design, too, but the more bird-hipped construction evolved several times; in the ornithischians, the therizinosaurs, and in birds.

Many scientists like phylogenetic systematics because it is rigorous, mechanical, and quantitative.[37] And that is true, up to a point. Before you can really do any phylogenetic analysis, however, you first have to code your characters. That is, you have to decide which character traits you are going to use as data, and

---

[37] Sober (1988) offers an excellent introduction to phylogenetic reasoning. Gee (2000) offers an especially (I would say, overly) enthusiastic account of how cladistic methods could make paleontology more rigorous.

then for each specimen in your set, you have to enter a value for each trait. So, for example, if you are studying hip joints, bird-hipped *vs.* lizard-hipped could be one character. Ultimately, a character could be any aspect of a dinosaur skeleton that is readily measurable: length of the skull, number of digits on the forelimbs, or whatever. But in order to get started at all, there are a couple of complex decisions that have to get made: Which specimens to use, and which characters to consider in the analysis. Things can get tricky when you have incomplete remains. Suppose, for example, that you have some ambiguous dinosaur skulls from way back in the Triassic, but you are missing the post-cranial skeletons. So you can't really tell whether the skulls came from bird-hipped or lizard-hipped animals. Once you have all of your characters coded, then you can use statistical analyses to determine which hypothesis about phylogenetic relationships is best supported. You can, in other words, crank out a result. However, we shouldn't be fooled by statistical rigor, as there is still an element of judgment and even connoisseurship that goes into the coding of characters. And it turns out that decisions about how to code characters can make a big difference to the results. And this, I will argue, is where aesthetic considerations sneak into play.

To provide just a little more context for understanding the debate about dinosaur phylogeny, we should think a little bit more about what Triassic dinosaurs were like. Dinosaurs first evolved and became well established during the Triassic. However, virtually all of the familiar, iconic, household name dinosaurs – from *Stegosaurus* to *Triceratops*, to *Tyrannosaurs* and *Argentinosaurus* – lived later, during the Jurassic and Cretaceous periods. During the Triassic, dinosaurs were less extravagant, and they were all rather similar to each other; mostly smallish to medium-sized animals, mostly bipedal. If we could time travel far enough back into the Triassic, the ancestors of *Triceratops* would actually look quite a bit like the ancestors of *T. rex*. This is a point that people routinely make about us mammals; we all evolved from boring, cat-sized nocturnal animals of the Mesozoic. Something analogous is true of dinosaurs. To make matters even worse, back in the Triassic, there were lots of other dinosaur-like reptiles running around – animals such as the silesaurs. So just figuring out whether something is an early dinosaur can be challenging, and may sometimes involve a paleontological judgment call. And to top it all off, the fossil record from the early to mid-Triassic is not so great. So from a purely epistemic point of view, it is not so easy to work out what was happening early on in the dinosaurs' evolutionary history.

It did not take long for critics to weigh in on Baron, Norman, and Barrett's proposed revision of dinosaur phylogeny. Some months later, the journal

Nature – where the original research had appeared – published a rejoinder by Langer, et al. (2017) as well as a response from Baron, Norman, and Barrett (2017b). Langer and colleagues expressed doubts about how Baron, Norman, and Barrett were coding characters:

> Our main concern is that the authors [i.e. Baron, Norman and Barrett (2017a)] were able to personally study fewer than half of the taxa in their analysis; the others were scored mostly based on published literature, which is problematic, because many characters relate to fine anatomical details, requiring first-hand study to be reliably documented (Langer et al. 2017, p. E1).

This is a fascinating sort of criticism to make; the worry is that the researchers were not themselves making a "first-hand study" of all the fossils in their dataset. Moreover, when the Langer team re-coded all the characters and ran the phylogenetic analysis again, they found support for the traditional ornithischian/saurischian phylogeny. "Character scoring changes explain our different results" (Langer et al. 2017, p. E1). In their own reply to the Langer team, Baron, Norman, and Barrett (2017b) say that "we disagree with many of the re-scorings" (p. E4). So it would seem that a massive disagreement about dinosaur phylogeny – is *T. rex* more closely related to *Triceratops* or to *Argentinosaurus*? – boils down to subtle disagreements about how to measure or describe the features of fossil dinosaur bones.

The main suggestion I wish to make here is that coding of characters – the "first-hand study" of fossils that serves as the starting point for any phylogenetic analysis – is in some degree an aesthetic activity. Of course, it is an epistemic activity too, as part of the point is to generate data for use as inputs to phylogenetic analysis. But the scoring of characters is at the same time a form of aesthetic engagement with fossils. It involves careful descriptive highlighting of some features while ignoring others, and it relies on a highly specific descriptive vocabulary. Phylogenetic reconstruction ultimately depends on perceptual engagement with fossils by experts who bring extensive background knowledge, as well as past experience with other fossils, to bear. The resulting phylogenetic analysis, in turn, can make a difference to our perception of the fossils in question, because the analysis can help reveal what's at stake descriptively.

To make this all a little more concrete, consider an unrelated description of a fossil, taken from a recent journal article. Here the scientists are describing the skull of a prehistoric lizard, *Elgaria panamintina*. The researchers explicitly say that the goal of their work is to provide a description for "future uses in morphological and phylogenetic studies of both extant species and fossils" (Ledesma and Scarpetta 2018, abstract). Here is just one bit of their description, quoted at some length so you can get a feel for the style:

The skull of *Elgaria panamintina* resembles the typical anguid condition of
being narrowed and elongated in the preorbital and postorbital regions of the
skull. The supratemporal fenestra is open but mediolaterally compressed, and
is bounded laterally by a complete postorbital and supratemporal bar. Cranial
osteoderms are fused to the midline roofing bones including the nasals, the
fused frontals, and the anterior portion of the parietal (Ledesma and Scarpetta
2018).

Now imagine pages and pages of this sort of description. This is a little bit like
reading a description of a fine wine. Ledesma and Scarpetta are highlighting
certain traits as being important – such as, for example, the "mediolateral
compression" of the supratemporal fenestra. (A "fenestra," by the way, is just
a hole in the skull.) The technical language makes it possible to draw extra-
ordinarily fine-grained morphological comparisons, even between different
particular specimens. Indeed, this is surely a case where possession of technical
descriptive language actually enhances one's perceptual abilities. It's easier to
see things that you have terms to describe. In this particular case, the scientists
were also using CT scanners to generate images of internal structure, and then
offering careful, precise descriptions of features that might not otherwise be
visible at all. The type of technical description has always been central to
paleontological practice. The construction of such descriptions is also a form
of aesthetic engagement, involving a sort of connoisseurship familiar from other
areas of art criticism. If this suggestion is anywhere near right, then it could turn
out that phylogenetic reconstruction ultimately depends on aesthetic engagement with fossils.

To return to the thread of our story, it turns out that Langer et al. (2017) and
Baron, Norman, and Barrett (2017a, 2017b) disagree significantly in their
characterization of a single specimen, known as *Pisanosaurus*, found in
Argentina in the 1960s. Much depends on how you characterize this one specimen. We know that *Pisanosaurus* was from the Triassic, around 220 million
years ago, a key moment in early dinosaur evolution. But the skeleton is so
incomplete that experts have disagreed about what it is. For example, we
possess the lower jaw, but none of the rest of the skull, which means that
we're missing lots of potentially diagnostic characters. Nor do we have the
complete hip joint. *Pisanosaurus* might be a really early ornithischian dinosaur,
but it might also be a silesaurid, closely related to dinosaurs but not belonging to
the dinosaur clade at all. Bonaparte (1976) thought the former, Angolin and
Rozadilla (2017) think the latter). Baron, Norman, and Barrett (2017b) write
that "Re-scoring of *Pisanosaurus* alone, based upon our personal observations
of the material, results not only in the recovery of Ornithoscelida, but also in the
identification of this enigmatic taxon as a silesaurid" (p. E4). That is, they count

it as a silesaurid, and when they do, they get the result that theropods form a clade with the ornithiscian dinosaurs. But if you include *Pisanosaurus* as an early ornithischian, that changes the entire phylogeny, and you get the result that theropods were more closely related to sauropods. Surprising as it sounds, some of the biggest questions of all about dinosaur phylogeny depend (at least for the time being) on different connoisseurs' rival interpretations of a single incomplete skeleton.

It is difficult to say how this debate about dinosaur phylogeny will play out. Seeing as how a single specimen, such as *Pisanosaurus*, can make such a difference to the phylogenetic analysis, it is entirely possible that finding new fossils from the Triassic could tip the evidential balance. From an epistemic point of view, this case might seem somewhat humbling. It would seem that dinosaur phylogeny is a lot more confusing and complicated than traditionally thought. However, this case also has an aesthetic side. The lesson is that you can't even do phylogenetic reconstruction without careful perceptual engagement with fossils – without treating the fossils as *aesthetic things* (section 7). Those *Pisanosaurus* fossils are precious because they are valuable epistemic tools that can be put to work in investigations of dinosaur phylogeny. But scientists' background interests in phylogenetic analysis, in turn, affect the practice of describing the fossil, and make a difference to which qualities and features are attended to. The phylogenetic disagreement is traceable, at least in part, to the fact that rival connoisseurs are giving different fine-grained descriptions of the same fragmentary fossil specimen. You can find aesthetic practices right at the heart of things, even where paleontological research might seem most machine-like and routinized.

9 Why are Dinosaurs Always Fighting?

In developing paleoaesthetics up to this point, I have stressed the ways in which the epistemic and aesthetic dimensions of scientific practice positively reinforce and mutually animate one another. For example, some aesthetic practices (such as field sketch work and fossil preparation) have contributed to the epistemic successes of the paleosciences. On the other hand, historical cognitivism implies that epistemic successes contribute in turn to deeper aesthetic appreciation and richer engagement with fossils and landscapes. But what about cases where deeply entrenched aesthetic biases seem to get in the way of good science? That such cases occur could be a source of resistance to my argument for the importance of the aesthetic dimension of paleontology.

There is a long tradition, in paleoart and museum exhibits, of showing *Tyrannosaurus* and *Triceratops* squaring off in one-on-one combat (Turner 2017b). This tradition goes back at least to 1906, when Charles Knight created an image of *Tyrannosaurus* menacing a couple of *Triceratops*. Knight repeated this theme in the famous and much reproduced mural that he created for the Field Museum of Natural History in 1928, but this time in the format of a one-on-one duel. The idea that *T. rex* fought *Triceratops* has had incredible staying power. Even at the height of the dinosaur renaissance, Robert Bakker's book, *The Dinosaur Heresies* (Bakker 1986), was published with a cover image of a theropod tussling with a ceratopsian (this time *Styracosaurus*). In other ways, the portrayal of the dinosaurs had changed completely. In Knight's work, the dinosaurs are so languid and slow that it could take all afternoon to see if *T. rex* gets anything to eat. But the animals on Bakker's cover are like action film dinosaurs, kicking up dust as they fight. It's quite remarkable that the massive revision of our view of dinosaurs that Bakker's book helped to consolidate made such little difference to the depiction of ritualized combat. It's hard to shake the feeling that in these images, the dinosaurs are acting out a familiar human script – it could be a boxing match, or a duel with pistols at high noon, or gladiatorial combat, or a light saber fight. Ritualized one-on-one violent combat is a human behavior that we seem to be projecting onto prehistoric nature, as if we're using the dinosaurs to show us something about ourselves.[38]

Incidentally, there is not a shred of evidence that *Tyrannosaurus* and *Triceratops* ever slogged it out in one-on-one combat. There are some suggestive toothmarks on the occasional *Triceratops* frill, but as the scientists who describe them point out, those could very easily have been made post-mortem (Fowler et al. 2012). Farke et al. (2009) did find some evidence of combat injuries in *Triceratops* skulls. But by comparing *Triceratops* with another ceratopsian – *Centrosaurus*, which lacked horns above the eyes – they disconfirmed the possibility that the injuries were caused by fighting with *T. rex*. The problem is that *Centrosaurus* also co-existed with *T. rex*, and the centrosaurs had much lower injury rates. If *Triceratops* was fighting with anyone, it was probably with members of its own species.

This case raises some questions traditionally associated with social constructivism (Hacking 2000). How might our own contingent cultural and aesthetic biases be affecting our efforts to reconstruct past life? What if we are representing dinosaurs in ways that just reflect our own aesthetic biases? One way that this can happen is that our biases can direct our attention. If we expect to see warfare and combat in nature, then we'll naturally get most excited about the animals with the most exaggerated weapons: the "dagger-like" teeth of *Tyrannosaurus*, the slashing claws of

---

[38] Fastovsky (2009) also explores the possibility that social and political context can influence our representations of dinosaurs.

Velociraptor, or that nasty cudgel on the tail of *Ankylosaurus*. As an example of how biases can direct our attention, consider the Farke et al. (2009) study mentioned above. That study is framed in a way that highlights the facial injuries in *Triceratops* specimens that the authors attribute to intraspecific horn locking. Just as interesting, though, is the suggestion that *Centrosaurus*, by contrast, did not engage in intraspecific combat! What makes the evidence of combat injuries a more important result than the lack of such evidence in a closely related lineage?

You might think that there are clear counterexamples to this suggestion that there is a cultural bias in favor of fighting dinosaurs, or dinosaurs with weapons. One of the most stunning finds of the dinosaur renaissance was Jack Horner's and Robert Makela's discovery of a dinosaur nesting ground in the late 1970s in Montana (Horner and Makela 1979). This was evidence of bird-like behaviour, dinosaurs flocking and nesting in colonies. The animals in question were Cretaceous hadrosaurs with no fancy weapons and no evidence of fighting. They didn't even have the cranial crests of some other better-known hadrosaurs. So much for the idea of fighting dinosaurs. In this case, however, Horner and Makela named the new genus *Maiasaura*, or "good mother lizard" – a pretty clear case of reading late twentieth-century gender norms back into the fossil record. Apparently the weaponless, communal, nesting dinosaurs must be "mothers."

In the film, *Jurassic Park*, one of the more bizarre and gratuitous deviations from scientific accuracy had to do with *Dilophosaurus*, an animal that plays a key role in the plot. On the big screen, *Dilophosaurus* is given a special weapon; it spits a gooey venom that contains a neurotoxin. There is not the slightest reason to think that the real animals were venomous. However, giving a formidable weapon to an otherwise less exciting dinosaur is a move that fits completely with audiences' expectation that dinosaurs were well-armed fighters, even though the move is in tension with the whole idea that dinosaurs were bird-like.

All of this points to a potential objection against paleoaesthetics. If aesthetic biases are such a problem, then perhaps it makes sense to exclude or externalize aesthetic values from the practice of paleontology. This might be one last reason for thinking the traditional view – the view that paleoscience should be thought of as a narrowly epistemic activity – was on the right track all along. The problem is that our aesthetic biases, such as our predilection for fighting dinosaurs, seem utterly divorced from any evidential considerations. So perhaps the right approach is just to try to keep those aesthetic values out of the science, as far as possible. Perhaps we should try to wall off our aesthetic interests from our epistemic interests, much as some think we should create a firewall between science and religion.

This externalizing move misdiagnoses the problem. Here it may help to borrow some ideas from Elizabeth Anderson (2004), who argues that one challenge for feminist philosophers of science who reject the value-free ideal is to distinguish legitimate from illegitimate ways in which non-epistemic values may influence scientific reasoning:

> We need to ensure that value judgments do not operate to drive inquiry to a predetermined conclusion. This is our fundamental criterion for distinguishing legitimate from illegitimate uses of values in science (Anderson 2004, p. 11).

The problem is not that we bring aesthetic values into play when trying to reconstruct dinosaurs. How could we do otherwise? The problem, following Anderson's suggestion, is that in this case our aesthetic predilections are driving inquiry "to a predetermined conclusion."

Nor is it helpful to argue that our preference for seeing dinosaurs fighting really only bears on paleoart and film – that is, on representations for broad nonscientific audiences. As Currie (2017b) has argued, paleoart contributes to the scientific process, even if the contribution is often indirect and roundabout. So, incidentally, does film. For example, some odd features of the Tyrannosaur's gait in the original *Jurassic Park* film, in the scene where it chases the speeding jeep, actually prompted questions that led to some insights on the part of researchers trying to estimate *T. rex* maximum running speeds (Turner 2009).

The problem is not that we bring aesthetic values into the process of science; the problem, rather, is that the specific values we bring are sometimes dumb and uninformed, and they then drive investigation in unhelpful ways. One takeaway message from the foregoing, especially Section 4, is that our aesthetic preferences are malleable. Everyone loves charismatic megafauna, but a little exposure and scientific knowledge can also turn one into a botanist or an entomologist. As soon as you recognize it for what it is, our cultural preoccupation with fighting dinosaurs begins to seem both silly and sad, a reflection of our culture's violent tendencies. Historical cognitivism implies that science itself can play a role in reshaping our aesthetic lives. As we learn more about the history of life on Earth, that changes how we engage with fossils and landscapes: better science means better aesthetic engagement.

Sometimes, the most productive response to aesthetic bias itself takes an artistic form. For example, Mark Witton (2015) explores other possibilities for bringing *Tyrannosaurus* and *Triceratops* together in a single frame. We know that in modern birds, interspecies adoption sometimes occurs. For example, brown-headed cowbirds will lay their eggs surreptitiously in the nests of other species, who then raise the cowbird hatchlings as their own. Did dinosaurs ever

Figure 13 *Triceratops* and *Tyrannosaurus* ignoring each other. Artwork by the author. This sketch is loosely inspired by John Conway's "Sleeping Tyrannosaurus rex," in Conway, Kosemen, and Naish (2012). But I added a Triceratops to make the point that the animals could have co-existed without fighting.

do this? We don't really know. But Witton points out that an image of a grown *T. rex* looking after a newly hatched *Triceratops* is not much more speculative than an image of the two facing off like gladiators. If the problem is that paleoart sometimes reinforces existing biases, then the solution might be more (and more imaginative) paleoart. The renderings of Conway, Kosemen, and Naish (2012) are a brilliant example of this. Figure 13 is another admittedly amateurish effort in this direction; a late Cretaceous scene in which *Triceratops* and *T. rex* are simply ignoring each other.

Another approach is to use what opportunities we have to shift attention away from the charismatic megafauna. One could, for example, lavish a bit more attention on fossils that might at first seem less exciting – like the remains of a freshwater clam bed that once lived in the shifting bend of a Cretaceous river, 70 million years ago.

10 Conclusion

Paleontology, like any natural science, is a set of investigative practices, with epistemic goals and tools. Paleontologists are trying to get knowledge of the deep past. And most of the work done so far in the philosophy of paleontology has focused narrowly – too narrowly, I argue – on this epistemic dimension of the science. I've argued that we need to expand our view of the science to include the aesthetic dimensions of its practice. This is because the aesthetic and epistemic dimensions of paleontological practice interact, mutually inform, and feed back on one another in a variety of complicated ways. Sometimes these two aspects of scientific practice may also blur together. One way to see this is to think of fossils as both epistemic tools and aesthetic things (Section 7). Bryan Norton and Sahotra Sarkar's account of transformative value gives us another way of thinking about epistemic-aesthetic feedback effects, and their account applies smoothly to fossils and landscapes (Section 4).

Paleontology has distinctively aesthetic as well as epistemic goals; paleontologists are not axiological obligates. There may be others – we should be pluralistic about aesthetic goods and values – but the two aesthetic aims that I've highlighted here are cultivation of sense of place, and better appreciation of fossils. Historical cognitivism shows how the epistemic practices associated with reconstructing the deep past contribute to these aesthetic aims (Section 3). The more you learn, say, about the fossilized clams in the Alberta badlands, the better you appreciate their qualities as aesthetic things *in situ*, and the deeper your connection to the place. On this view, functional morphology also turns out to be a form of aesthetic investigation (Section 5). All of this is to say that the investigative practices of paleontology have aesthetic payoff.

Turning things around the other way, other paleoaesthetic practices – practices that are or look a lot like art – also have investigative payoff. We cannot fully appreciate the epistemic successes of the paleosciences by focusing solely on the epistemic dimensions of the practice (Section 6). We also need to consider things like fossil preparation, paleoart, 3D printing, and the creation of geological field sketches. None of these practices, taken on their own terms, fit the old picture of disembodied scientific agents generating and testing hypotheses, and all of them borrow from the arts. Not only that, but even investigative practices that do seem purely epistemic – things like phylogenetic reconstruction – sometimes depend crucially on researchers' aesthetic engagement with particular fossils, as well as careful morphological observation and description (Section 8). The latter involves a kind of aesthetic connoisseurship that's reminiscent of other forms of art criticism.

Along the way, I've also identified some factors that have made the aesthetic dimensions of the paleosciences harder to see, or difficult to acknowledge. One problem is that theory-centric philosophy of science frames the issues in an unhelpful way, treating aesthetic values as candidate criteria of theory choice, thus subordinating them completely to epistemic concerns. Another issue is that certain metaphors – especially the textual metaphor of the fossil "record" – lock us into seeing fossils in narrow epistemic terms, as an evidence base (Section 7). A third serious issue is that cultural/aesthetic biases, such as our predilection for seeing dinosaurs fighting with each other (Section 9), can get in the way of good investigation. I've also shown how to address these three issues: first, by adopting a more practice-oriented approach; second, by considering alternative metaphors; and third, by using some of the other lessons of paleoaesthetics to help correct unhelpful aesthetic biases.

I have been developing these arguments from a fairly high-altitude perspective, and no doubt many details need more attention. Nevertheless, paleoaesthetics has some important consequences for ongoing debates in the philosophy of science. First, my hope is that it might offer something of a vindication of practice-oriented approaches to the philosophy of science, by showing how such approaches can yield a richer understanding of how science works. Second, paleoaesthetics promises to expand the discussion of values in science, a discussion that has, for contingent reasons, focused more narrowly on inductive risk, and on ethical, social, and political values that come into play in policy-relevant science. Third, paleoaesthetics has consequences for the scientific realism debate, especially for the traditional realist project of explaining the success of science (Section 6). If I am right, the successes that need explaining include aesthetic successes. And attending to the aesthetic dimensions of the scientific practice can generate better explanations of the epistemic successes too.

Finally, in developing paleoaesthetics, I have wanted to give a fuller philosophical account of the joyful enchantment that paleoscience affords. Perhaps this also provides a new way of thinking about how science more generally connects us not only with places and with fossils, but with other living things and with the natural systems that we all depend on.

References

Agnolin, F.L. and Rozadilla, S. (2017). Phylogenetic reassessment of *Pisanosaurus mertii*, Casamiquela, 1967, a basal dinosauriform from the late Triassic of Argentina. *Journal of Systematic Paleontology* 16(10): 853–879. DOI:10.1080/14772019.2017.1352623

Anderson, E. (2004). Uses of value judgments in science: A general argument, with lessons from a case study of feminist research on divorce. *Hypatia* 19(1): 1–24.

Arkhipkin, A.I. (2014). Getting hooked: The role of a U-shaped body chamber in adult heteromorph ammonites. *Journal of Molluscan Studies* **80**, 354–364.

Bakker, R.T. (1986). *The Dinosaur Heresies: New theories unlocking the mystery of dinosaurs and their extinction*. New York: William Morrow & Co.

Baron, M.G., Norman D.B., and Barrett, P.M. (2017a). A new hypothesis of dinosaur relationships and early dinosaur evolution. *Nature* **543**, 501–506. (2017b). Barrett et al. reply. *Nature* **551**, E4-E5.

Basso, K.H. (1996). Wisdom sits in places: Notes on a Western Apache landscape. In S. Feld and K.H. Basso, eds., *Senses of Place*. Santa Fe, NM: School of American Research Press, pp. 53–90.

Beatty, J. (2016). What are narratives good for? *Studies in History and Philosophy of Biological and Biomedical Sciences* **58**, 33–40.

Berleant, A. (1995). *The Aesthetics of Environment*. Philadelphia, PA: Temple University Press.

Bokulich, A. (2018). Using models to correct data: Paleodiversity and the fossil record. *Synthese* https://doi.org/10.1007/s11229-018-1820-x.

Bonaparte, J.F. (1976). *Pisanosaurus mertii* Casamiquela 1967 and the origin of the *Ornithischia*. *Journal of Palaeontology* **50**, 808–820.

Boyd, R. (1990). Realism, approximate truth, and method. *Minnesota Studies in the Philosophy of Science* **14**, 355–391.
(1984). The current status of scientific realism. In J. Leplin, ed., *Scientific Realism*. Berkeley, CA: University of California Press, pp. 41–82.

Boym, S. (2002). *The Future of Nostalgia*. New York: Basic Books.

Bradley, R. (2000). *An Archaeology of Natural Places*. London: Routledge.

Brady, E. (2011). The ugly truth: Negative aesthetics and environment. *Royal Institute of Philosophy Supplement* **69**, 83–99.
(2003). *Aesthetics of the Natural Environment*. Edinburgh, UK: University of Edinburgh Press.

Bruchac, M. (2005), Earthshapers and placemakers: Algonkian Indian stories and the landscape. In M. Wobst and C. Smith, eds., *Indigenous Archaeologies: Politics and practice.* London: Routledge, pp. 56–80.

Carlson, A. (2009). *Nature and Landscape: An introduction to environmental aesthetics.* New York: Columbia University Press.

(2008). Nature and positive aesthetics. In A. Carlson and S. Lintott, eds., *Nature, Aesthetics, and Environmentalism: From beauty to duty.* New York: Columbia University Press, pp. 211–237.

(2000). *Aesthetics and the Environment: The appreciation of nature, art, and architecture.* London: Routledge.

(1981). Nature, aesthetic judgment, and objectivity. *Journal of Aesthetics and Art Criticism* **40**(1), 15–27.

(1977). Appreciation and the natural environment. *Journal of Aesthetics and Art Criticism* **37**(3), 267–275.

Chang, H. (2014). Epistemic activities and systems of practice: Units of analysis in philosophy of science after the practice turn. In L. Soler, S. Zwart, M. Lynch, and V. Israel-Jost, eds., *Science after the Turn to Practice in the History, Philosophy, and Social Studies of Science.* New York: Routledge, pp. 69–79.

Chapman, R. and Wylie, A. (2016). *Evidential Reasoning in Archaeology.* London: Bloomsbury Press.

Cleland, C. (2011). Prediction and explanation in historical natural science. *British Journal for the Philosophy of Science* **62**(3), 551–582.

(2002). Methodological and epistemic differences between historical science and experimental science. *Philosophy of Science* **69**(3), 474–496.

Conway, J., Kosemen, C.M., and Naish, D. (2012). *All Yesterdays: Unique and speculative views of dinosaurs and other prehistoric animals.* Irregular Books.

Cunningham, J.A. Rahman, I.A., Lautenschlager, S., et al. (2014). A virtual world of paleontology. *Trends in Ecology and Evolution* **29**(6), 347–357.

Currie, A. (2018). *Rock, Bone, and Ruin: An optimist's guide to the historical sciences.* Cambridge, MA: MIT Press.

(2017a). Hot-blooded gluttons: Dependency, coherency, and method in the historical sciences. *British Journal for the Philosophy of Science* **68**(4), 929–952.

(2017b). Paleoart as science, *Extinct: The philosophy of palaeontology blog.* February 27, 2017. www.extinctblog.org/extinct/2017/2/27/paleoart-as-science.

(2017c). The secret epistemology of paleontological fieldwork. Extinct: The philosophy of palaeontology blog. September 4, 2017. www.extinctblog.org /extinct/2017/9/4/the-secret-epistemology-of-paleontological-fieldwork.

(2015). Marsupial lions and methodological omnivory: Function, success, and reconstruction in paleobiology. *Biology and Philosophy* **30**(2), 187–209.

(2012). Convergence, contingency, and morphospace. Review of G.R. McGhee. *Convergent Evolution: Limited Forms Most Beautiful.* *Biology and Philosophy* 27: 583–593.

Currie, A. and Sterelny, K. (2017). In defence of story-telling. *Studies in History and Philosophy of Science* **62**, 14–21.

Currie, A. and Turner, D. (2016). Introduction: Scientific knowledge of the deep past. *Studies in History and Philosophy of Science* **55**, 43–46.

Darwin, C. (1853/1964). *On the Origin of Species.* A Facsimile of the First Edition. Cambridge, MA: Harvard University Press.

Dawkins, R. (1998). *Unweaving the Rainbow: Science, delusion, and the appetite for wonder.* New York: Houghton Mifflin.

Dietl, G.P. and Flessa, K.W. (2011). Conservation paleobiology: Putting the dead to work. *Trends in Ecology and Evolution* **26**(1), 30–37.

Douglas, H.E. (2009). *Science, Policy, and the Value-Free Ideal.* Pittsburgh, PA: University of Pittsburgh Press.

(2000). Inductive risk and values in science. *Philosophy of Science* **67**(4), 559–579.

Elgin, C.Z. (2002). Art in the advancement of understanding. *American Philosophical Quarterly* 39(1): 1–12.

Elliot, R. (1982). Faking nature. *Inquiry* **26**(1), 81–93.

Elliott, K.C. (2017). *A Tapestry of Values: An introduction to values in science.* Oxford, UK: Oxford University Press.

Engler, G. (1990). Aesthetics in science and art. *British Journal of Aesthetics*, 24–33.

Farke, A.A., Wolff, E.D.S., and Tanke, D.H. (2009). Evidence of combat in Triceratops. *PloS One* **4**(1), e4252.

Fastovsky, D.E. (2009). Ideas in dinosaur paleontology: resonating to social and political context. In D. Sepkoski and M. Ruse, eds., *The Paleobiological Revolution: Essays on the growth of modern paleontology.* Chicago, Il: University of Chicago Press, pp. 239–253.

Finkelman, L. (2017). Stop the clocks (and other geological timescale metaphors, too)! *Extinct: The philosophy of palaeontology blog* 23, October, 2017. www.extinctblog.org/extinct/2017/10/23/stop-the-clocks-and-the-other-geologic-timescale-metaphors-too

Forber, P. and Griffith, E. (2011). Historical reconstruction: Gaining access to the deep past. *Philosophy, Theory, and Practice in Biology* 3(3). https://quod.lib.umich.edu/cgi/t/text/text-idx?cc=ptb;c=ptb;c=ptbbio; idno=6959004.0003.003;rgn=main;view=text;xc=1;g=ptbbiog.

Fowler, D., Scannella, J.B., Goodwin, M.G., and Horner, J.R. (2012). How to eat a Triceratops: Large sample of toothmarks provides new insight into the feeding behavior of Tyrannosaurus. *Journal of Vertebrate Paleontology* 32(5, abstracts volume), 96.

Frodeman, R. (2014). Hermeneutics in the field: The philosophy of geology. In D. Ginev, ed., *The Multidimensionality of Hermeneutic Phenomenology*. Dordrecht: Springer, pp. 69–79.

(2003). *Geo-Logic: Breaking ground between philosophy and the earth sciences*. New York: State University of New York Press.

Galison, P. (1987). *How Experiments End*. Chicago, IL: University of Chicago Press.

Gee, H. (2000). *In Search of Deep Time: Beyond the fossil record to a new history of life*. Ithaca, NY: Cornell University Press.

Gell, A. (1998). *Art and Agency: An anthropological theory*. Oxford, UK: Oxford University Press.

Gilbert, G.K. (1877). *Report on the Geology of the Henry Mountains*. US Geological Survey Unnumbered Series. Washington, DC: US Government Printing Office.

Godfrey-Smith, P. (1994). A modern history theory of functions. *Nous* **28**, 344–362.

Godlovitch, S. (2008). Icebreakers: Environmentalism and natural aesthetics. In A. Carlson and S. Lintott, eds. *Nature, Aesthetics, and Environmentalism: From beauty to duty*. New York: Columbia University Press, pp. 133–150.

Gould, S.J. (1988). *Time's Arrow, Time's Cycle: Myth and metaphor in the discovery of geological time*. Cambridge, MA: Harvard University Press.

Gould, S.J. and Lewontin, R. (1979). The spandrels of San Marco and the panglossian paradigm: A critique of the adaptationist programme. *Proceedings of the Royal Society B* **205**, 581–598.

Hacking, I. (2000). *The Social Construction of What?* Cambridge, MA: Havard University Press.

Hacking, I. (1983). *Representing and Intervening: Introductory Topics in the Philosophy of Natural Science*. Cambridge, UK: Cambridge University Press.

Heringman, N. (2004). *Romantic Rocks, Aesthetic Geology*. Ithaca, NY: Cornell University Press.

Hesse, M. (1966). *Models and Analogies in Science*. South Bend, IN: Notre Dame University Press.

Hettinger, N. (2008). Objectivity in environmental aesthetics and protection of the environment. In A. Carlson and S. Lintott, eds., *Nature, Aesthetics, and Environmentalism: From beauty to duty*. New York: Columbia University Press, pp. 413–438.

Hitchcock, E. (1858/1974). *A Report on the Sandstone of the Connecticut Valley, Especially Its Fossil Footmarks*. New York: Arno Press.

Horner, J.R. and Makela, R. (1979). Nest of juveniles provides evidence of family structure among dinosaurs. *Nature* **282**, 296–298.

Hume, D. (1988). *Dialogues Concerning Natural Religion*, second edition, R. Popkin, ed. Indianapolis, IN: Hackett Publications.

Ivanova, M. (2017). Aesthetic values in science. *Philosophy Compass* **12**(10), e12433.

James, S. (2015). Why old things matter. *Journal of Moral Philosophy* **12**(3), 313–329.

Jeffares, B. (2008). Testing times: Regularities in the historical sciences. *Studies in History and Philosophy of Biological and Biomedical Sciences* **39**(4), 469–475.

Jones, N. (2012). Science in three dimensions: The print revolution. *Nature* **487**(7405), 22–23.

Korsmeyer, C. (2016). Real old things. *British Journal of Aesthetics* **56**(3), 219–231.

   (2012). Touch and the experience of the genuine. *British Journal of Aesthetics* **52**(4), 365–377.

Kosso, P. (2001). *Knowing the Past: Philosophical issues of history and archaeology*. San Jose, CA: Humanity Books.

Landman, N.H., Kruta, I., Denton, J.S.S., and Cochran, J.K. (2014). Getting unhooked: Comment on the hypothesis that heteromorph ammonites were attached to kelp branches on the sea floor. *Journal of Molluscan Studies* **82**, 351–356.

Langer, M.C. et al. (2017). Untangling the dinosaur family tree. *Nature* **551**, E1-E3.

Langer, S.K. (1942). *Philosophy in a New Key*. Cambridge, MA: Harvard University Press.

Larson, B. (2014). *Metaphors for Environmental Sustainability*. New Haven, CT: Yale University Press.

Laudan, L. (2004). The epistemic, the cognitive, and the social. In P. Machamer and G. Wolters, eds., *Science, Values, and Objectivity*. Pittsburgh, PA: University of Pittsburgh Press, pp. 14–23.

(1981). A confutation of convergent realism. *Philosophy of Science* **48**(1), 19–49.

Ledesma, D.T. and Scarpetta, S.G. (2018). The skull of the gherronotine lizard *Elgaria panamintina* (Squamata: Anguidae). *PLoS ONE* **13**(6), e0199584.

Leopold, A. (1989). *A Sand County Almanac*. Oxford: Oxford University Press.

Leplin, J. (1997). *A Novel Defense of Scientific Realism*. Oxford: Oxford University Press.

Lipton, P. (2004). *Inference to the Best Explanation*, second edition. London: Routledge.

Longino, H. (1990). *Science as Social Knowledge: Values and objectivity in scientific inquiry*. Princeton NJ: Princeton University Press.

Lyell, C. (1853). *Principles of Geology*, ninth edition. New York: D. Appleton & Co.

Maclaurin, J. (2003). The good, the bad and the impossible: A critical notice of "theoretical morphology: the concept and its applications" by George McGhee. *Biology and Philosophy* 18:463–476

MacLaurin, J. and Sterelny, K. (2008). *What is Biodiversity?* Chicago, IL: University of Chicago Press.

Matthes, E.H. (2018). Authenticity and the aesthetic experience of history. *Analysis*, any028, https://doi.org/10.1093/analys/any028

(2017). Palmyra's ruins can rebuild our relationship with history. *Aeon Magazine* (4 September, 2017). https://aeon.co/ideas/palmyras-ruins-can-rebuild-our-relationship-with-history

Matthews, P. (2008). Scientific knowledge and aesthetic appreciation of nature. In A. Carlson and S. Lintott, eds., *Nature, Aesthetics, and Environmentalism: From beauty to duty*. New York: Columbia University Press, pp. 188–204.

McAllister, J.W. (1989). Truth and beauty in scientific reason. *Synthese* **78**(1), 25–51.

McGhee, G. (2011). *Convergent Evolution: Limited forms most beautiful*. Cambridge, MA: MIT Press

McPhee, J. (2000). *Annals of the Former World*. New York: Farrar, Strauss, & Giroux.

McShea, D.W. and Brandon, R.N. (2010). *Biology's First law: The tendency for diversity and complexity to increase in evolutionary systems*. Chicago IL: University of Chicago Press.

Millikan, R.G. (1984). *Language, Thought, and Other Biological Categories*. Cambridge, MA: MIT Press.

(1989). In defense of proper functions. *Philosophy of Science* **56**, 283–302.

Mitchell, W.J.T. (1998). *The Last Dinosaur Book: The life and times of a cultural icon*. Chicago, IL: University of Chicago Press.

Miyake, T. (2018). Scientific realism and the earth sciences. In J. Saatsi, ed., *The Routledge Handbook of Scientific Realism*. London: Routledge, pp. 333–344.

Monks, N., and P. Palmer (2002). *Ammonites*. Washington, DC: Smithsonian Books.

Monnet, C., Klug, C., and De Baets, K. (2015). Evolutionary patterns of Ammonoids: Phenotypic trends, convergence, and parallel evolution. In C. Klug, ed., *Ammonoid Paleobiology: From macroevolution to paleogeography*. Dordrecht: Springer, pp. 95–142.

Montgomery, D.R. (2012). *The Rocks Don't Lie: A geologist investigates Noah's flood*. New York: W.W. Norton.

Mustoe, G.E (2001). Enigmatic origin of ferruginous "coprolites:" Evidence from the Miocene Wilkes Formation, southwestern Washington. *GSA Bulletin* **113**(6), 673–681.

Neander, K. (1991). Functions as selected effects: The conceptual analyst's defense. *Philosophy of Science* **56**, 168–184.

Norton, B.G. (1987). *Why Preserve Natural Variety?* Princeton, NJ: Princeton University Press.

Nussbaum, M. (2003). *Upheavals of Thought: The intelligence of emotions*. Cambridge: Cambridge University Press.

Oldroyd, D.R. (2013). Maps as pictures or diagrams: The early development of geological maps. In V.R. Baker, ed., *Rethinking the Fabric of Geology*. Boulder, CO: Geological Society of America, Special Paper 502, pp. 41–102.

  (2006). *Earth Cycles: A historical approach*. Westport, CT: Greenwood Press.

O'Loughlin, I. and McCallum, K. (2019). The aesthetics of theory selection and the logics of art. *Philosophy of Science* **86**(2), 325–343.

Parsons, G. (2008a). *Aesthetics and Nature*. New York: Continuum.

  (2008b). Nature appreciation, science, and positive aesthetics. In A. Carlson and S. Lintott, eds., *Nature, Aesthetics, and Environmentalism: From beauty to duty*. New York; Columbia University Press, pp. 302–324.

Parsons, G. and Carlson, A. (2008). *Functional Beauty*. Oxford, UK: Oxford University Press.

Parsons, K.M. (2001). *Drawing Out Leviathan: Dinosaurs and the science wars*. Bloomington, IN: Indiana University Press.

Paul, L.A. (2015). *Transformative Experience*. Oxford, UK: Oxford University Press.

Philpotts, A.R. and Asher, P.M. (1994). Magmatic flow-direction indicators in a giant diabase feeder dike. *Connecticut Geology* **22**, 363–366.

Potochnik, A. (2015). The diverse aims of science. *Studies in History and Philosophy of Science* **53**, 71–80.

Preston, B. (1998). Why is a wing like a spoon? A pluralist theory of function. *Journal of Philosophy* **115**, 215–54.

Psillos, S. (2018). The realist turn in the philosophy of science. In J. Saatsi, ed., *The Routledge Handbook of Scientific Realism*. New York: Routledge, pp. 20–34.

   (1999). *Scientific Realism: How science tracks truth*. London: Routledge.

Putnam, H. (1978). *Meaning and the Moral Sciences*. London: Routledge and Kegan Paul.

Raab, T. and Frodeman, R. (2002). What is it like to be a geologist? A phenomenology of geology and its epistemological implications. *Philosophy and Geography* **5**(1), 69–81.

Rahman, I.A., Adcock, K., and Garwood, R.J. (2012). Virtual fossils: A new resource for science communication in paleontology. *Evolution: Education and Outreach* **5**(4), 635–641.

Raup, D.M. (1966). Geometric analysis of shell coiling: General problems. *Journal of Paleontology* **40**, 1178–1190.

Raymo, C. and Raymo, M. (2001). *Written in Stone: A geological history of the Northeastern United States*. Hensonville, NY: Black Dome Press.

Rheinberger, H.-J. (1997). *Toward a History of Epistemic Things: Synthesizing proteins in the test tube*. Palo Alto, CA: Stanford University Press.

Rieppel, L. (2016). Casting authenticity. *Extinct: The philosophy of palaeontology blog*. (1 February, 2016). www.extinctblog.org/extinct/2016/1/28/cast ing-authenticity

   (2015). Prospecting for dinosaurs on the mining frontier: The value of information in America's gilded age. *Social Studies of Science* **45**(2), 161–186.

Rolston, H., III. (1989). *Philosophy Gone Wild*. Buffalo, NY: Prometheus Books.

Rooney, P. (1992). On values in science: Is the epistemic/non epistemic distinction useful? *PSA: Proceedings of the Biennial Meeting of the Philosophy of Science Association 1992*, pp. 13–22.

Rossetter, T. (2018). Realism on the rocks: Novel success and James Hutton's theory of the earth. *Studies in History and Philosophy of Science* **67**, 1–13.

Rouse, J. (2002). *How Scientific Practices Matter: Reclaiming philosophical naturalism*. Chicago, IL: University of Chicago Press.

(2015). *Articulating the World: Conceptual understanding and the scientific image*. Chicago: IL: University of Chicago Press.

Rudwick, M.J.S. (2016). *Earth's Deep History: How it was discovered and why it matters*. Chicago, IL: University of Chicago Press.

(1972). *The Meaning of Fossils: Episodes in the history of palaeontology*. Chicago, IL: University of Chicago Press.

Ruse, M. (2013). *The Gaia Hypothesis: Science on a pagan planet*. Chicago, IL: University of Chicago Press.

(2005). Darwinism and mechanism: metaphor in science. *Studies in History and Philosophy of Biology and Biomedical Sciences* **36**(2), 285–302.

Russow, L.-M. (1981). Why do species matter? *Environmental Ethics* **3**(2), 101–112.

Saatsi, J. (2018). Introduction: Scientific realism in the 21st century. In J. Saatsi, ed., *The Routledge Handbook of Scientific Realism*. New York: Routledge, pp. 1–4.

Saito, Y. (2008). Appreciating nature in its own terms. In A. Carlson and S. Lintott, eds., *Nature, Aesthetics, and Environment: From beauty to duty*. New York: Columbia University Press, pp. 151–168.

Sandis. C. (2016). An honest display of fakery: Replicas and the role of museums. *Royal Institute of Philosophy Supplement* **79**, 241–259.

Sarkar, S. (2005). *Biodiversity and Environmental Philosophy*. Cambridge: Cambridge University Press.

Scannella, J. and Horner, J.R. (2010). *Torosaurus* Marsh, 1891, is Triceratops Marsh 1889 (*Ceratopsidae: Chasmosaurinae*): Synonymy through ontogeny. *Journal of Vertebrate Paleontology* **30**(4), 1157–1168.

Schindler, S. (2018). *Theoretical Virtues in Science: Uncovering reality through theory*. Cambridge: Cambridge University Press.

Seilacher, A. (1988). Why are Nautiloid and Ammonite sutures so different? *Neues Jahrbuch für Geologie und Paläontologie* **177**, 41–69.

Sepkoski, D. (2012). *Rereading the Fossil Record: The growth of paleobiology as an evolutionary discipline*. Chicago, IL: University of Chicago Press.

Sepkoski, D. and Ruse, M. (2009). *The Paleobiological Revolution: Essays on the growth of modern paleontology*. Chicago, IL: University of Chicago Press.

Sharwood, S. (2014). Metre-long dinosaur poo going under the hammer. *The Register* (25 July, 2014). www.theregister.co.uk/2014/07/25/metrelong_dinosaur_poo_going_under_the_hammer/

Smith, J.E.H. (2019) Beginning at the beginning: Leibniz and the texts of deep history. Extinct: The philosophy of palaeontology blog (18 February, 2019). Available online at: www.extinctblog.org /extinct/2019/2/18/beginning-at-the-beginning-leibniz-and-the-texts-of-deep-history.

Sober, E. 1988. *Reconstructing the Past: Parsimony, evolution, and inference.* Cambridge, MA: MIT Press.

(1986). Philosophical problems for environmentalism. In B. Norton, ed., *The Preservation of Species: The value of biological diversity.* Princeton, NJ: Princeton University Press, pp. 173–195.

Sober, E. and M. Steel. (2014). Time and knowability in evolutionary processes. *Philosophy of Science* **81**(4), 558–579.

Spencer, P. (1993). The coprolites that aren't: The straight poop on specimens from the Miocene of southwestern Washington State. *Ichnos* **2**(3), 231–236.

Steel, D. (2010). Epistemic values and the argument from inductive risk. *Philosophy of Science* 77(1): 14–34.

Switek, B. (2014). Was six-million-year-old turd auctioned for $10,000 a faux poo? *National Geographic* (28 July, 2014). https:// news.nationalgeographic.com/news/2014/07/140729-dinosaur-coprolite -paleontology-dung-fossil-auction/

(2010). *Written In Stone: Evolution, the Fossil Record, and Our Place in Nature.* Bellevue Library Press.

Tamborini, M. (2019). Technoscientific approaches to deep time. *Studies in History and Philosophy of Science.* Published online 8 March, 2019. https://doi.org/10.1016/j.shpsa.2019.03.002.

Tucker, A. (2011). Historical science, over- and under-determined: A study of Darwin's inference of origin. *British Journal for the Philosophy of Science* **62**(4), 805–829.

(2004). *Our Knowledge of the Past: A philosophy of historiography.* Cambridge, UK: Cambridge University Press.

Turner, D.D. (2018). Three kinds of realism about historical science. In J. Saatsi, ed., *The Routledge Companion to Scientific Realism.* New York: Routledge, pp. 321–332.

(2017a). Science, religion, and bad poetry. Extinct: The philosophy of palaeontology Blog. 2 January, 2017. www.extinctblog.org/extinct/2016/ 12/24/science-religion-and-bad-poetry.

(2017b). From the war of nature. *Extinct: The philosophy of palaeontology Blog.* 8 May, 2017. Available online at www.extinctblog.org/extinct/2017/ 5/2/from-the-war-of-nature

(2016). Conservation Paleobiology. *Extinct: The philosophy of palaeontology Blog.* 21 March, 2016. Available online at www.extinctblog.org /extinct/2016/3/16/conservation.

(2014). Philosophical issues in recent paleontology. *Philosophy Compass* **9**, 494–505.

(2011). *Paleontology: A philosophical introduction.* Cambridge: Cambridge University Press.

(2009). Beyond detective work: Empirical testing in paleobiology. In Sepkoski, D., and M. Ruse, eds., *The Paleobiological Revolution: Essays on the growth of modern paleontology.* Chicago, IL: University of Chicago Press, pp. 201–214.

(2007). *Making Prehistory: Historical science and the scientific realism debate.* Cambridge: Cambridge University Press.

(2005a). Are we at war with nature? *Environmental Values* **14**, 21–36.

(2005b). Local underdetermination in historical science. *Philosophy of Science* **72**(1), 209–230.

(2000). The functions of fossils: Inference and explanation in functional morphology. *Studies in History and Philosophy of Biology and Biomedical Sciences* **31**, 193–212.

Uno, H. (1998). Chemical conviction: Dickinson, Hitchcock, and the poetry of science. *The Emily Dickinson Journal* **7**(2), 95–111.

Van Fraassen, B.C. (1980). *The Scientific Image.* Oxford UK: Oxford University Press.

Walsh, K. (2019), Newton's Scaffolding: The instrumental roles of his optical hypotheses. In A. Vanzo and P. Anstey, eds., *Experiment, Speculation and Religion in Early Modern Philosophy*, London:Routledge.

Ward, P. (1981). Shell sculpture as defensive adaptation in Ammonoids. *Paleobiology* **7**, 96–100.

Waters, C.K. (2014). Shifting attention from theory to practice in philosophy of biology. In M.C. Galavotti, D. Dieks, W.J. Gonzalez, S. Hartmann, T. Uebel, and M. Weber eds., *New Directions in the Philosophy of Science*, Berlin: Springer International Publishing, pp. 121–139

(2008). Beyond theoretical reduction and layer-cake antireduction: How DNA. In Michael Ruse, ed., *Oxford Handbook to the Philosophy of Biology.* New York: Oxford University Press, pp. 238–26

Witton, M. (2015). Tyrannosaurus and Triceratops – friends at last? Markwitton .com blog (25 March 2015). Available online at http://markwitton-com.blogspot.com/2015/03/tyrannosaurus-and-triceratops-friends.html

Witton, M.P., Naish, D., and Conway, J. (2014). State of the paleoart. *Palaeontologia Electronica* **17**(3), 5E. https://doi.org/10.26879/145

Woody, A.I. (2014). Chemistry's periodic law: Rethinking representation and explanation after the turn to practice. In *Science after the Practice Turn in the Philosophy, History, and Social Studies of Science*, L. Soler, S. Zwart, M. Lynch, and V. Israel-Jost, eds., New York: Routledge, pp. 123–150.

Wray, K. Brad. (2018). Success of science as a motivation for realism. In J. Saatsi, ed., *The Routledge Handbook of Scientific Realism*. New York: Routledge, pp. 37–46.

Wright, L. (1973). Functions. *Philosophical Review* **82**, 139–168.

Wylie, A. (2002). *Thinking from Things: Essays in the philosophy of archaeology*. Berkeley, CA: University of California Press.

(1999). Rethinking unity as a working hypothesis for philosophy of science: How archaeologists exploit the disunity of science, *Perspectives on Science* 7.3 (1999): 293–317.

Wylie, C.D. (2015). The artist's piece is already in the stone: Constructing creativity in paleontology laboratories. *Social Studies of Science* **45**(1), 31–55.

(2009). Preparation in action: Paleontological skill and the role of the fossil preparator. In: M.A. Brown, J.F. Kane, and W.G. Parker, eds., *Methods in Fossil Preparation: Proceedings of the first annual fossil preparation and collections*.

Acknowledgments

Portions of this work have been shared in the form of talks at the University of Durham; at Wesleyan University; at the 2018 Society for Philosophy of Science in Practice meeting in Ghent, Belgium; and at a once-in-a-lifetime retreat at Dinosaur Provincial Park in Alberta, Canada, in August 2017. Many thanks to Don Brinkman for showing some of us philosophers around the Canadian badlands.

Adrian Currie provided incredibly helpful and detailed comments on the full manuscript. I am also deeply grateful to Grant Ramsey and Michael Ruse for their support for this project. Thanks especially to Ruse for his kind help, advice, and encouragement over the years.

I am grateful beyond words to my dear friends and fellow Extinct bloggers: Leonard Finkelman, Joyce Havstad, and Adrian Currie. I'm not sure if they know how much they have affected my intellectual life for the better. Many of the ideas that I explore here were first floated as trial balloons on the blog, and appear here in heavily reworked and expanded form.

In the spring of 2017, I had an opportunity to spend a Fulbright semester at the University of Calgary, Canada. The philosophy department at Calgary was a wonderful place to work, and I owe so much to conversations there with Marc Ereshefsky, Ken Waters, and many others: Soohyun Ahn, Justin Caouette, Megan Delahanty, David Dick, Brian Hanley, Noa Latham, Oliver Lean, Ann Levey, Alison McConwell, Mark Migotti, Celso Neto, T.J. Perkins, Lauren Ross, Josh Stein, and Nicole Wyatt. The Calgary paleontologists – Jason Pardo, Larry Powell, Jessica Theodor, Selena Robson, and others – also graciously allowed me to crash their weekly reading group. I am deeply grateful to Fulbright Canada for this support.

I have the amazing good fortune to be married to an archaeologist, and Michelle Turner has introduced me to landscape phenomenology, materiality theory, and other ideas that have changed my thinking about paleontology. Some of these ideas were inspired by dinner conversations with Michelle and her 2016 field crew in New Mexico (Max Forton, Josh Jones, Randy McGuire, Lubna Omar, Sam Stansel, Kellam Throgmorton, and Ruth Van Dyke).

At Connecticut College, I'm lucky to have some terrific colleagues who like discussing philosophy, field science, and environmental issues: Robert Askins, Lindsay Crawford, Jane Dawson, Simon Feldman, Michelle Neely, Jennifer Pagach, Kristin Pfefferkorn, Doug Thompson, Larry Vogel, Mel Woody, and others. In the fall of 2017, I inflicted an early draft of this essay on my philosophy of science students at Connecticut College, and the manuscript is much improved on account of their tough criticisms.

I've been helped along by conversations with many others, including Kelli Carlson, Carol Cleland, Craig Fox, Rob Inkpen, Erynn Johnson, Elizabeth Jones, Teru Miyake, Rune Nyrup, Thomas Rossetter, Joe Rouse, Elise Springer, Alison Wylie, and above all Caitlin Wylie, whose work on fossil preparators has inspired much of what follows. I apologize to those whom I'm sure I've forgotten to mention here.

This Element is dedicated to my mother, Mary Turner, who during much of my childhood worked as the curator of a decorative arts museum. I spent lots of time hanging out in the museum with the Wedgwood porcelain, cut glass, and other breakable things. She went on to work for many years with the Illinois Historic Preservation Agency. Thanks to her, my younger life was full of serious reflection on our relationship to the past. And it's been a blessing to discuss this project with her too, from time to time.

Cambridge Elements ☰

Philosophy of Biology

Grant Ramsey
KU Leuven

Grant Ramsey is a BOFZAP Research Professor at the Institute of Philosophy, KU Leuven, Belgium. His work centers on philosophical problems at the foundation of evolutionary biology. He has been awarded the Popper Prize twice for his work in this area. He also publishes in the philosophy of animal behavior, human nature, and the moral emotions. He runs the Ramsey Lab (theramseylab.org), a highly collaborative research group focused on issues in the philosophy of the life sciences.

Michael Ruse
Florida State University

Michael Ruse is the Lucyle T. Werkmeister Professor of Philosophy and the Director of the Program in the History and Philosophy of Science at Florida State University. He is Professor Emeritus at the University of Guelph, in Ontario, Canada. He is a former Guggenheim fellow and Gifford lecturer. He is the author or editor of over sixty books, most recently *Darwinism as Religion: What Literature Tells Us about Evolution; On Purpose; The Problem of War: Darwinism, Christianity, and their Battle to Understand Human Conflict;* and *A Meaning to Life.*

About the Series

This Cambridge Elements series provides concise and structured introductions to all of the central topics in the philosophy of biology. Contributors to the series are cutting-edge researchers who offer balanced, comprehensive coverage of multiple perspectives, while also developing new ideas and arguments from a unique viewpoint.

Philosophy of Biology

Elements in the Series

Printed in the United States
By Bookmasters